LEG AND HOOF CARE FOR HORSES

KNACK™

LEG AND HOOF CARE FOR HORSES

A Complete Illustrated Guide

MICAELA MYERS

Photographs by Kelly Meadows

Veterinary review by Linda Byer, DVM, MS

Guilford, Connecticut
An imprint of The Globe Pequot Press

Text design by Paul Beatrice

Library of Congress Cataloging-in-Publication Data is available on file.

ISBN 978-1-59921-396-5

Printed in China
10 9 8 7 6 5 4 3 2 1

Front cover photo credit: Kelly Meadows
Back cover photo credit: Moira C. Harris
All interior photos by Kelly Meadows except for:
Photos pp. ix, x (bottom), xi, 23 (right), 33 (right), 35 (right), 49 (left), 52 (left), 55 (right), 56 (left), 58 (right), 62 (left), 64 (left), 65 (left), 67 (left), 68, 69 (right), 73 (left), 75 (left), 76 (left), 78, 82 (left), 85 (right), 89 (right), 91, 94 (right), 104 (right), 114 (left), 116 (left), 122 (left), 142 (right), 145 (left), 154 (right), 155 (left), 157 (right), 158 (right), 159 (left), 160 (right), 162, 163 (right), 168 (right), 170, 179 (left), 183 (left), 191 (right), 198 (right), 200 (right), 208 (left), 209 (left), 216 (left), 225 (top), 226 (top), 229 (bottom) by Moira C. Harris
Photo p. 27 (left) courtesy of Lisa Carter, LAMP, Heavenly Gaits Equine Massage (Hico, TX)
Photos pp. 67 (right), 102 (right) by Jessie Shiers
Photo p. 87 (left) courtesy of EQUIOXX
Photo p. 142 (left) by Fran Jurga/Hoofcare & Lameness Journal
Photos pp. 145 (right), 157 (left), 166 (left) by Micaela Myers
Photos pp. 149 (left), 150 courtesy of Richard Stephenson, Pool House Equine Clinic
Photos pp. xii, 1 (left), 7 (left), 15, 16 (right), 17 (left), 77, 80-81, 87, 92-93, 95 (right), 97 (right), 98 (left), 100 (right), 115 (left), 116 (right), 117 (left), 128 (right), 130 (right), 131, 132 (right), 134 (right), 135, 136 (left), 137 (left), 140 (right), 141 (right), 152 (right), 153 (left), 157 (left), 204 (left), 205 (right), 206 (left) courtesy of Sylvia Greenman, DVM, Inc.
Photos pp. 171 (right), 181 (left) by Kelsey Hanrahan
Photo p. 185 (left) by Jason Smith
Illustrations pp. 220-23 © American Quarter Horse Association and reprinted with permission

This book is dedicated to my pony Blackjack, who was a wonderful teacher and a hard working friend; and to my dad, Don Hanley, for all of his help and support over the years—from mucking corrals and building shelters to renovating horse trailers.

Special Thanks

First and foremost, thank you to all the veterinary researchers helping us to better understand and treat lameness—without their work this book wouldn't be possible. Secondly, thanks to my horses, my farriers, and my vet, Dr. Linda Byer, who have taught me so much over the years—together we've managed navicular disease, laminitis, abscesses, ringbone, sidebone, and more. Thanks also go to my friend Moira C. Harris and my husband, Joe Myers, for their support, and to my mother, Anne Hanley, who has read everything I've ever written and helped me care for my horses throughout the years. Thanks to my friend Dana Willis for always lending a helping hand and to my friend and fellow author "Dr." Toni McAllister for her help with this book. Last but not least, special thanks to Sylvia Greenman, DVM, for providing many of the x-ray, ultrasound, and nuclear scintigraphy images in this book.

CONTENTS

INTRODUCTION

We've all heard the adage "no hooves, no horse," and I would say that "no legs, no horse" is just as true. Whether you're a reiner, jumper, trail rider, dressage enthusiast, barrel racer, or anything in between, your horse's future depends largely on how his hooves and legs hold up. Our role as riders and horse owners has a big impact on our horses' current and future soundness.

I've owned horses for twenty-five years, and the value of healthy legs and hooves hit home for me at a young age. When I was just ten years old, I slipped the bareback pad on and tried to get my Welsh pony, Blackjack, to trot around the field. She kept refusing—she didn't even want to walk and kept turning back toward her corral. The more insistent I got, the more persistent she got. Pretty soon she did trot—right back to her pen where she promptly jumped into the water trough! I sat there on her back with her front feet in the water wondering what on earth was going on. It turns out Blackjack was suffering from laminitis. Her hooves were hot and sore, and she was a lot smarter than I was!

Unfortunately, our learning curve was slow, and that first bout of laminitis was not her last. After her third incidence of laminitis, we'd finally learned the correct maintenance strategies to keep her healthy. Luckily, being the tough pony she was, she lived happily into her late twenties.

From trail riding Blackjack and showing her in gymkhana events and pleasure driving, I went on to compete in western pleasure and trail, English pleasure and hunter hack, and barrel racing. I also tried cutting and jumping. My partners were Bear, a huge, gentle bulldog Quarter Horse with tiny feet, who suffered from navicular disease; Kahn, a sweet, playful Arabian who constantly gave himself splints and other interference injuries; Berry, a Quarter Pony who developed ringbone and navicular syndrome due to pacing; and Sonny, a small, feisty Quarter Horse mare whose upright pasterns and hooves led to sidebone. While confronting these issues with the help of my vet and farrier taught me a lot, I certainly wish I had the knowledge then that I have now. While invaluable, generous experts like vets and farriers are an integral part of your team, their job description doesn't include being a reference book, and time constraints and hectic schedules prohibit them from giving each horse owner a comprehensive lesson on every issue that arises, so the horse owner also needs to be proactive in learning about legs and hooves.

The past twenty-five years of horse ownership reinforced for me how precious soundness is. I also learned that no horse—regardless of his breed or use—is immune to leg and hoof issues. My horses also taught me how important the owner's role is in both prevention and management.

There are horse owners who start out like I did, taking soundness for granted. These owners usually end up learning the hard way that legs and hooves need careful care. There are other owners who have experienced

soundness issues first hand and hope to avoid ever meeting them again. They search elusively for a horse that will show up 100 percent clean on the prepurchase examination, meaning the vet finds not one issue or blemish during the exam or on x-ray. A horse like this is rare, and rather than searching for elusive perfection, it's important to know which conformation faults may predispose a horse to problems, which veterinary findings are merely cos-

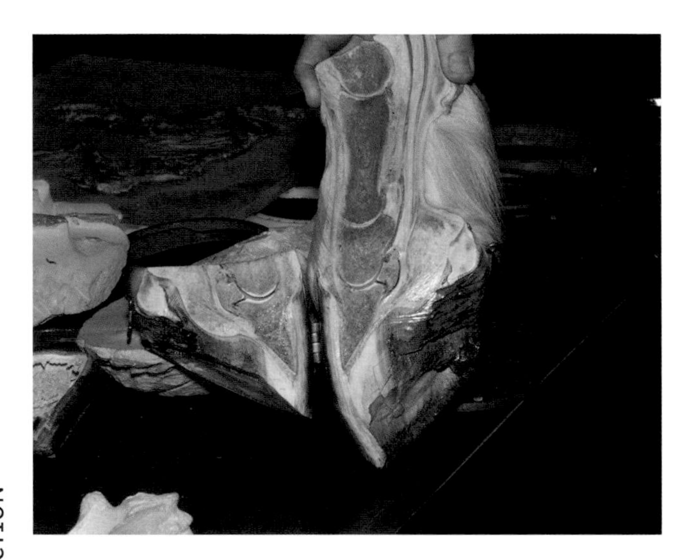

metic, which issues you can live with and which findings may rule a horse out as an athletic partner. It's also important to know that a horse that's "clean" and sound now won't necessarily be sound long-term or even tomorrow! The best you can do is become an educated horse owner who takes maintenance and prevention seriously and understands that even when issues arise, oftentimes the horse can be returned to activity with the right treatment and care. This book will help you become that educated horse owner.

It was with these ideas and knowledge that I set about researching and writing this book. It's my intention for *Knack Leg and Hoof Care for Horses* to help you:

Understand a horse's hooves and legs. The average horse is one thousand pounds or more, but his legs are only the width of our own legs in places. This book starts with a look inside the horse's amazing legs and hooves so that their inner workings are no longer a mystery. It will also teach you how conformation affects soundness and movement. Understanding the legs and hooves will help you better understand the problems that can affect them.

Learn how to prevent hoof and leg problems. Don't be one of those horse owners who wait to give hooves and legs the attention they deserve until it's too late. By following the prevention strategies outlined in this book, you and your horse are likely to have many more active years together.

Learn key care strategies. This book will help you master wrapping techniques (including standing wraps, sweats and poultices), clear up skin problems and combat pests, learn when to use heat or cold therapy, learn how to feed a balanced diet, know how to properly stock your first-aid kit, and much more.

Understand and recognize hoof and leg problems. It's safe to say that every horse will suffer one or more leg or hoof issues in his lifetime. Being able to recognize problems early is imperative for a positive outcome, and properly understanding issues will help you during every phase of treatment.

Become knowledgeable about treatment options for hoof and leg problems. Knowing what options are available and their pros and cons can help you make informed decisions about your horse's care and maintenance. New treatments are constantly on the horizon, and one may be right for you and your horse.

What sets *Knack Leg and Hoof Care for Horses* apart from other books are its hundreds of color photographs and its clear, concise, easy to reference format. With this book, there's no need to read through pages and pages to find the answers you seek. We've made sure every sentence counts so that you can find the information you need quickly. We've also made sure it's reader friendly and not written in language that requires a medical degree to decipher. Sure, the Internet is a great resource (and many

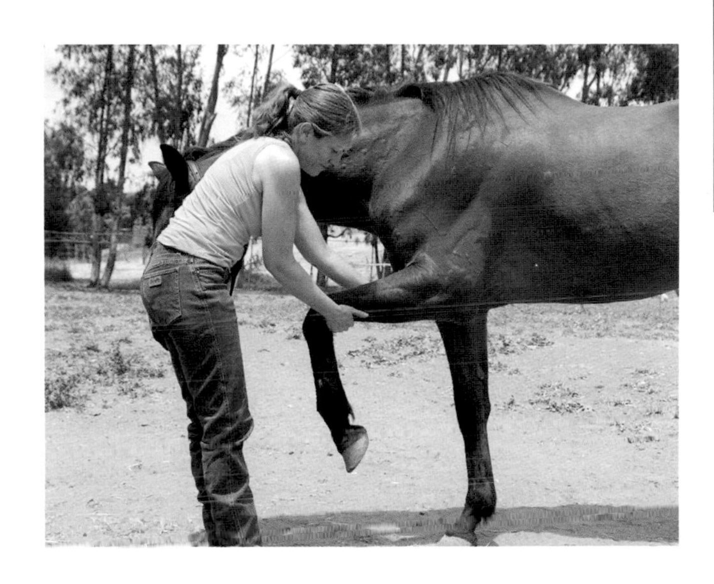

wonderful sites are listed in the Resources section of this book), but chances are you'll have to visit many sites and scan countless pages before you find what you're looking for. Here, the answers are at your fingertips—no searching, no sifting. Included are entire chapters devoted to the common problems of navicular disease, laminitis, and degenerative Joint disease/arthritis—with the latest information. This comprehensive book also includes dozens of other leg and hoof problems that can affect your horse. If I'd only had this book twenty-five years ago!

With the information contained in this book, you'll be armed with the knowledge you need to care for your horse's legs and hooves, keeping your horse happy and healthy so that you both can enjoy your pursuits together for many, many years to come.

BONES OF THE HOOF

The short pastern bone, coffin bone, and navicular bone play key roles in your horse's hoof

Although you can't see most of them, the horse's hoof contains three bones that serve an important role and can also cause serious problems when things go awry. At the top of the hoof is the short (or lower) pastern bone (also called the middle or second phalanx). The short pastern bone sits halfway in and halfway out of the hoof, connecting with the long pastern above it. The bottom of the short pastern bone is rounded in shape and fits nicely into the socket shape created for it by the other bones inside the hoof.

The coffin bone (also known as the pedal bone or distal phalanx) sits below the short pastern within the hoof capsule. The coffin bone is almost hoof shaped itself and indeed

Short Pastern Bone

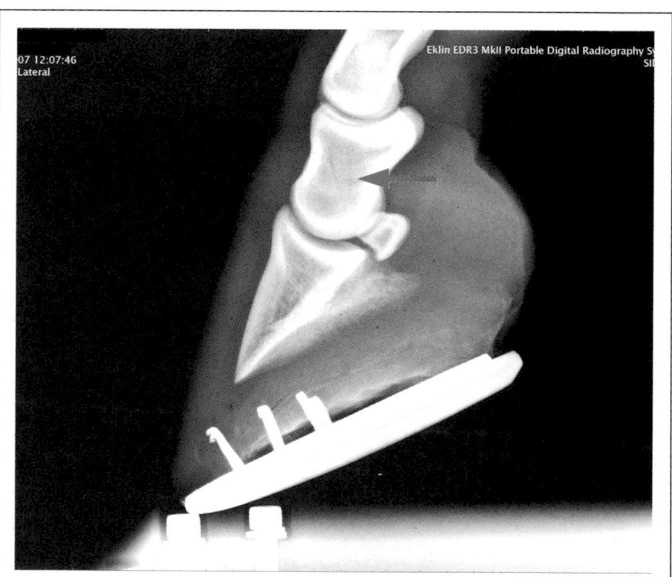

- The short pastern bone (shown in the photograph above) is so called because it's shorter than the long pastern bone above it.
- The pastern bones are integral players in shock absorption, which is why they sit at an angle rather than straight up and down, like the cannon bone (see page 14).
- While the short pastern bone doesn't shift a great deal, the long pastern bone above it increases its angle during movement.

Coffin Bone

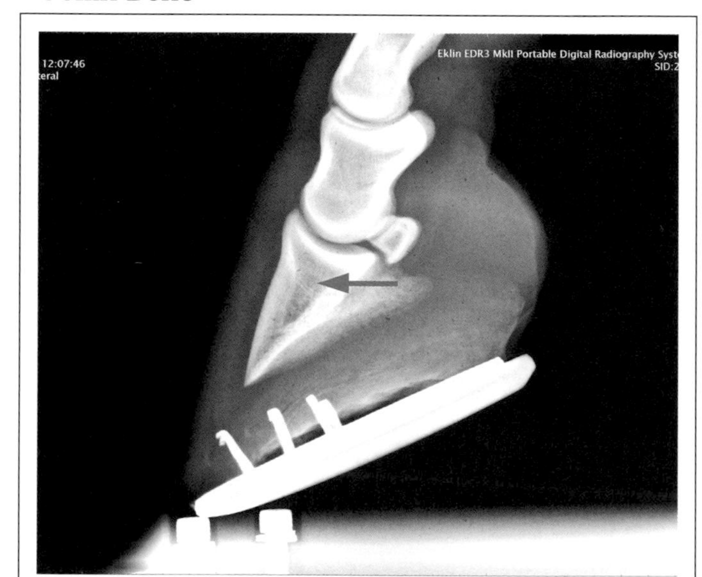

- The coffin bone (shown in this graphic) is rounded in front, much like a hoof.
- The underside of the coffin bone is slightly concave.
- The front upper tip of the coffin bone is called the extensor process and is the highest point of the coffin bone.
- The back left and right sides, or wings, are called the palmer processes.

dictates the natural size and shape of the horse's hoof. Just as you can visually see that the front hooves of a horse have a slightly different shape than the hind hooves (which are often narrower, with a more concave sole), the coffin bones of a horse also reflect these shapes internally.

Between the top, back part of the coffin bone and the lower rear side of the short pastern sits the tiny, relatively thin bone that's made such a big name for itself: the navicular bone (also known as the distal sesamoid bone). This precarious placement is what gives this little bone so many options for trouble. Between bones are joints, and the navicular bone contacts two other bones and, therefore, two joint surfaces. In addition, the important deep digital flexor tendon runs down the horse's leg and over the navicular bone to the coffin bone.

Navicular Bone Side View

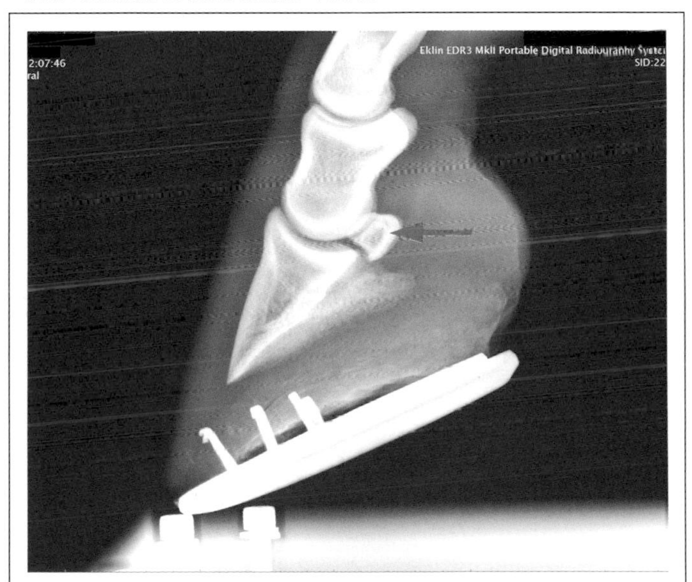

- The navicular bone looks like a small nugget from the side view (as shown in this graphic).
- Sesamoid bones, like the distal sesamoid (another name for navicular bone), are found where tendons change direction or pass over bony prominences.
- The navicular bone acts as a fulcrum or pulley for the deep digital flexor tendon, which runs down the horse's leg and over the navicular bone.

Navicular Bone Back View

- The navicular bone looks kind of like a whale's tail lifted out of the water when you view it from the back.
- The navicular bone is longer than it is thick.
- The palmer processes, or wings, of the coffin bone enclose the navicular bone from the sides.

SENSITIVE PART OF THE HOOF

Inside the hoof are living, sensitive parts that are vital to hoof health

We talked about the bones of the hoof, but what's between the bones and the hard hoof wall you see? This area is largely inhabited by the corium (or "quick")—connective tissue containing a vascular matrix of arteries. The corium is responsible for hoof growth, for connecting hoof to bone, and for providing nourishment to the hoof.

The coronary corium creates cells and hoof growth, much like the soft, sensitive parts beneath our fingernails create our relatively tough nails. The coronary corium is located beneath the coronary band/coronet—the soft area at the top of the hoof where skin and hair meet hoof. New hoof wall growth comes from this area, and the coronary corium constantly produces new cells. As older cells move along their journey, they harden and become the keratinized hoof wall that we're familiar with.

Much as hoof wall growth comes from the coronary corium,

Coronary Band, Coronet, Periople

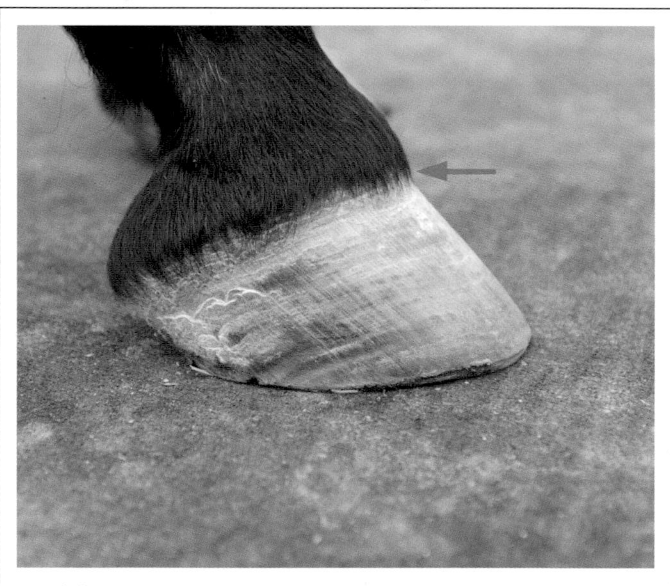

- While some consider the coronary band to encompass the area inward of the coronet, most people use the terms *coronet* and *coronary* band interchangeably.
- Inward and above the coronet is the perioplic corium, which distributes nourishment to the top of hoof.
- Around the top part of the hoof, the periople acts kind of like a cuticle with a waxy substance. As the hoof grows out, the old periople material dries and forms the outer layer of hoof wall called the stratum tectorium.

Laminar Corium

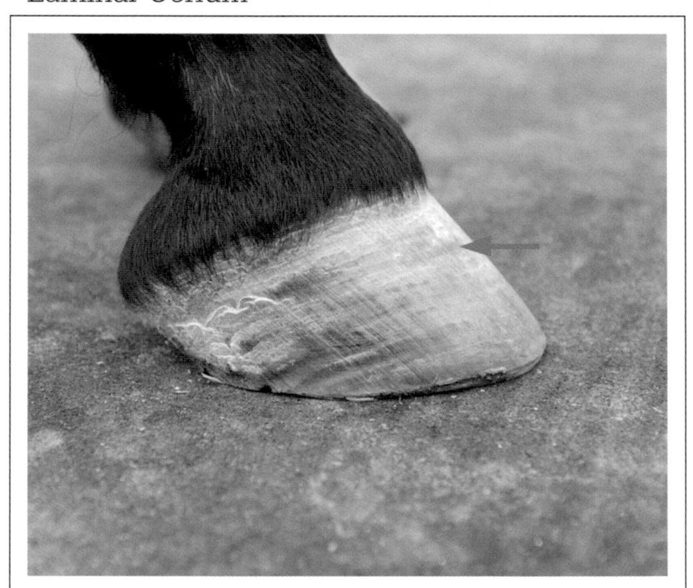

- The graphic on the photograph above shows the laminar corium inside the keratinized, or hard, hoof capsule.
- The epidermal (insensitive) laminae that line the hoof capsule and dermal (sensitive) laminae lining the coffin bone are like Velcro or fingers, connecting and holding on to one another to keep the coffin bone suspended within the hoof capsule.

sole tissues are created from corium of the sole. Frog tissues are likewise produced from frog corium beneath the frog.

The area where the wall and the sole attach is known as the white line. The white line is somewhat flexible. This thin band of connective tissue is lighter colored, hence the name "white" line, and can be seen where the wall and sole connect when you pick up a hoof and look at the bottom of it.

The laminar corium produces the laminae inside the hoof that have the important job of connecting the coffin bone to the hoof capsule. The dermal (sensitive) laminae—the laminae with blood and nerves—reach out from the coffin bone and connect with the hoof wall's epidermal laminae, like fingers. The epidermal laminae are lined with basement membrane—a sheet of connective tissue that helps bridge the dermal and epidermal laminae, playing a key role.

Frog Corium

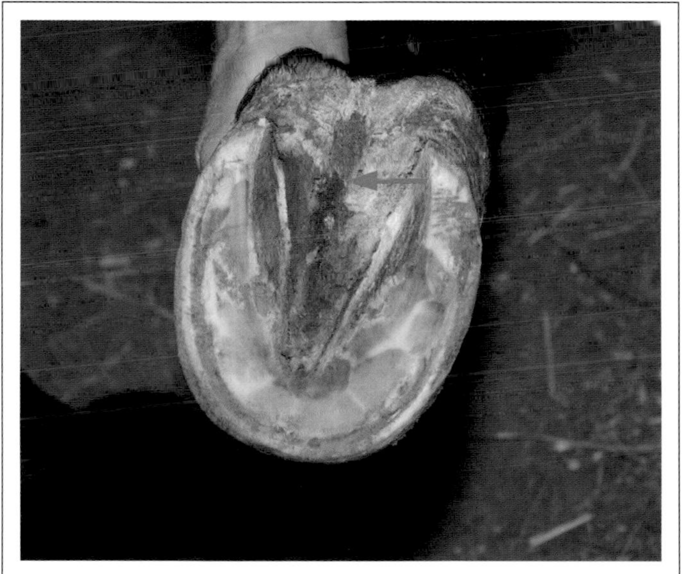

- The graphic on this photo shows the frog corium, which is responsible for creating the tissues that eventually make up the insensitive part of the frog that your hoof pick comes in contact with when you clean your horse's hooves.
- The frog is the triangular region in the middle, back part of the hoof.
- The frog is softer to the touch than the sole and thought to play a role in the horse's traction.

Solar Corium

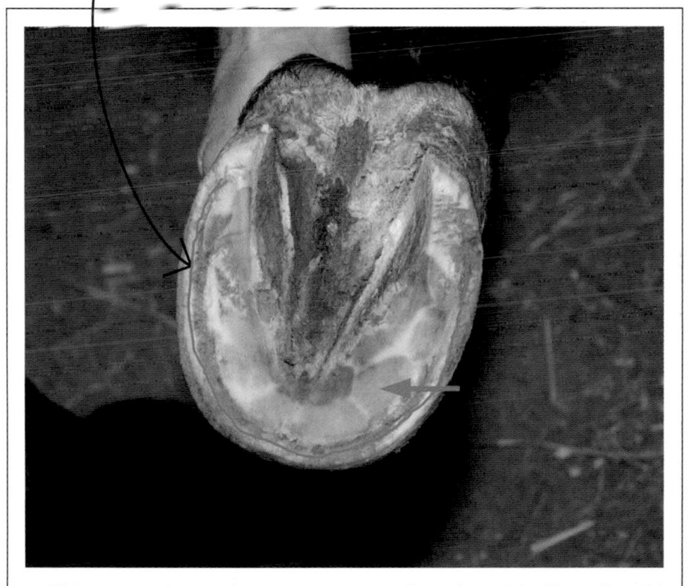

- This arrow shows the general area of the corium of the sole, or solar corium, from which the insensitive sole we see and touch is produced.
- The band between where the sole and wall connect, called the white line, should be uniform in width all the way around (not wide or stretched in places).
- Trouble in the white line area can indicate laminar problems higher up in the hoof.

INSENSITIVE PARTS OF THE HOOF

The harder, keratinized parts of the hoof include the hoof wall, sole, bars, and frog

The hard, keratinized (or cornified) parts of the hoof, also sometimes called the horn, are the result of hoof growth. These parts have no nerves or blood vessels—just like the tips of our fingernails that we trim off. These insensitive parts include the hoof wall; the sole, or underside of the horse's hoof; the frog, or outer layer of frog tissue; and the bars. The bars are at the back part of the hoof wall. They turn inward along the outside of each collateral groove. The collateral grooves are the deep indentations along either side of the frog.

The outermost layer of the hoof wall is called the pigmented layer. It gets its color or pigment from cells inside

The Hoof Wall

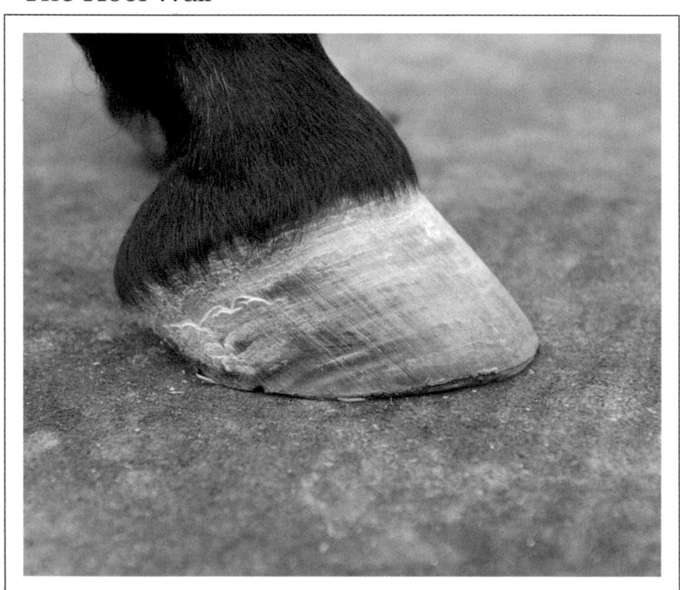

- The hoof wall is keratinized and super tough.
- Keratin is the protein that creates our fingernails and hair as well as the horn tissue that makes up the hoof wall.
- The hoof wall is actually a bunch of tiny hairlike keratinized tubules "glued" together.
- These tiny tubules give the hard hoof a bit of elasticity.

Hoof Wall Underside View

- When you pick up your horse's hoof, you can see how thick his hoof wall is.
- Everything from the white line out is hoof wall.
- Some horses have thinner hoof walls than others.
- Too thin of a hoof wall is considered a defect and can make shoeing difficult.

the coronary band, where growth begins. You'll notice that some horses with white hair above the hoof have light color hooves. However, it is a myth that white hooves are weaker than black hooves.

When you pick up the horse's hoof, you can see the hoof wall and inward of that the white line (line where sole and wall connect). The hoof wall is thicker toward the bottom of the hoof. The hoof wall grows approximately a quarter inch a month (slightly more in some horses), which is why horse's hooves need regular trimming.

However, not all these keratinized, hard surfaces are the same. The hoof wall is harder than the sole, and the frog is softer than the sole. Though not as tough, the frog is more elastic and the sole more supple than the hoof wall. The frog also has the highest moisture content of the three, at about 50 percent. In addition, it's easy to see and feel that the horse's heel and heel bulbs are likewise softer than the sides and front of the hoof wall.

Sole

- Even though the outer part of the sole that we see and touch when we pick up a horse's hoof is insensitive, just underneath it lies the corium and all the sensitive structures of the hoof. This is why nails cause a problem if they enter the sole or frog; yet, nails driven into the hoof wall do not.
- The equine sole is slightly concave rather than flat, although some horses are more flat-footed than others.

Bars

- The bars are the back part of the hoof wall, where the wall curves in along the outer side of the frog's collateral grooves.
- The bars should be about half the length of the frog.
- The bars are strong and work as braces, keeping the foot from overly contracting or expanding.
- Just like the rest of the hoof wall, the bars need regular trimming to maintain their proper form.

HOOF ELASTICITY

The hoof is designed to absorb shock and withstand weight-bearing.

With the average horse weighing around 1,000 pounds, it's not surprising that his hooves are designed with weight-bearing and shock absorption capacities. During the weight-bearing phase of movement, when the horse places his weight on the hoof, the hoof changes shape slightly to absorb this shock. Almost all parts of the hoof are capable of some type of elasticity, however minor or major, to adjust during this phase. Even the tough hoof wall has some elastic properties.

However, at the root of the horse's shock-absorbing mechanism is the digital cushion. The digital cushion is a fibrous wedge-shaped pad of connective tissue that sits above the

Digital Cushion Side View

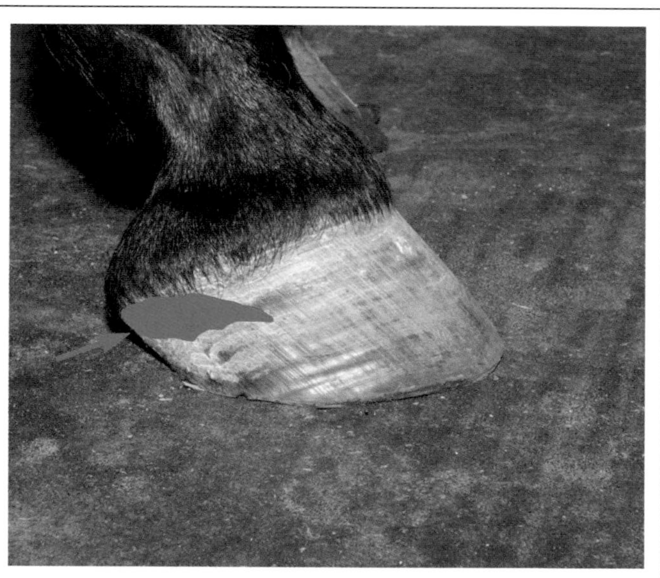

- As you can see from this graphic, the wedge-shaped digital cushion sits toward the back of the horse's hoof.
- The digital cushion is very elastic and helps to absorb shock.
- The digital cushion is "sandwiched" during the weight-bearing movement of the horse with pressure from the bones above and from the ground below.

Digital Cushion Underside View

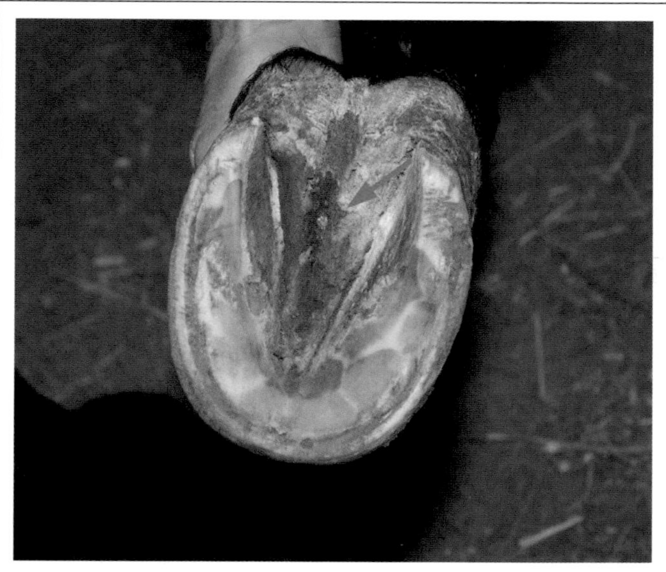

- The general area of the digital cushion is shown with this arrow from underneath the hoof.
- The digital cushion sits above the frog.
- The frog itself is also somewhat elastic.
- During the weight-bearing phase of movement, the digital cushion expands against the lateral cartilages on either side of it.

frog, taking up the majority of the space in this back portion of the hoof capsule. Directly above the digital cushion is the deep digital flexor tendon, the coffin bone, and the navicular bone. The rear portion of the digital cushion helps form the heel bulbs. During the weight-bearing phase of movement, the digital cushion expands and absorbs pressure from the coffin bone and short pastern above that.

On either side of the digital cushion sit the strong but also somewhat flexible lateral cartilages. The lateral cartilages help protect the sensitive tissues in between them.

ZOOM

The expanding and then contracting that take place in the hoof as the result of movement—walking, trotting, cantering, and so on—play a role in circulation by creating a pumping action. During the weight-bearing phase, blood is moved up the leg, and new blood returns as the weight is lifted.

Hoof Expansion Side View

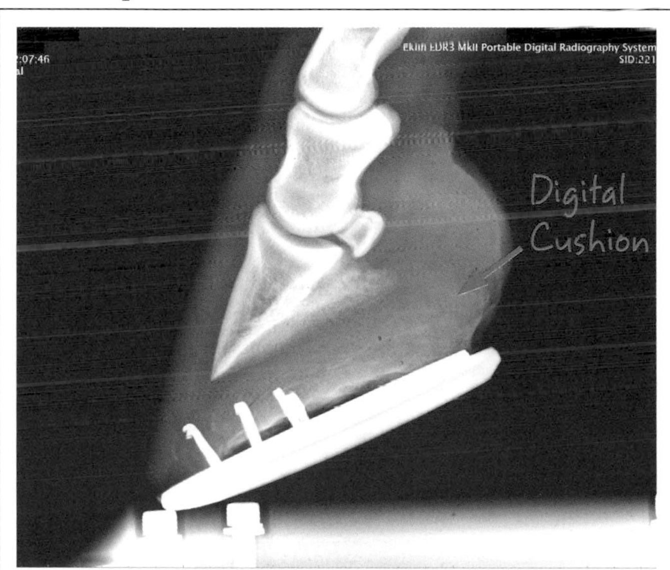

- During the weight-bearing phase, the coffin bone is pressed down, causing the digital cushion to flatten somewhat, and blood is moved out of the foot.
- The digital cushion and internal structures of the hoof play a key role.
- Imagine what would happen if the coffin bone made direct contact with the hoof capsule during weight bearing.

Hoof Expansion Underside View

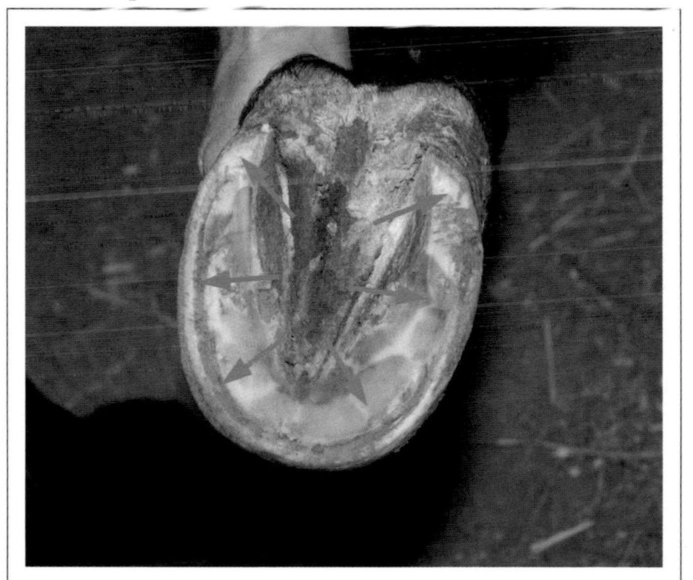

- These arrows show how even the "hard," keratinized parts of the hoof play a role in hoof expansion.
- During the weight-bearing phase, the sole loses some concavity and flattens slightly. This in turn also spreads the heels and expands the hoof wall and bars.
- The heel bulbs' ability to spread slightly during weight bearing plays a key role in hoof expansion.

HOOF AND PASTERN ANGLES

The angle of the hoof and pastern affects the horse's movement and soundness

Watch a horse walk, trot, canter, or jump, and you'll see his long pastern bone (which sits just above the short pastern bone) move the fetlock joint down toward the ground during the weight-bearing phase. This is part of the horse's shock-absorbing system and why the pastern bones sit at an angle rather than straight up and down like the cannon bone.

Not all pasterns, however, were created equal. Some pasterns are too long and sloping, making it appear as if the fetlock joint above the long pastern bone will fall all the way to the ground with each stride. This places added strain on the fetlock joint, tendons, and ligaments. A pastern bone that's short and too upright, on the other hand, creates a choppy

Correct Shoulder and Pastern Angle

- The graphics in the photograph above show a horse with a shoulder angle similar to the hoof and pastern angles.
- Ideally, you can draw a straight line up the front of your horse's hoof wall and his pastern, and the angle of that line will be similar to the angle of his shoulder blade.
- A horse with correct angles has a smoother gait than a horse with upright pasterns and shoulders.

Correct Pastern Angle

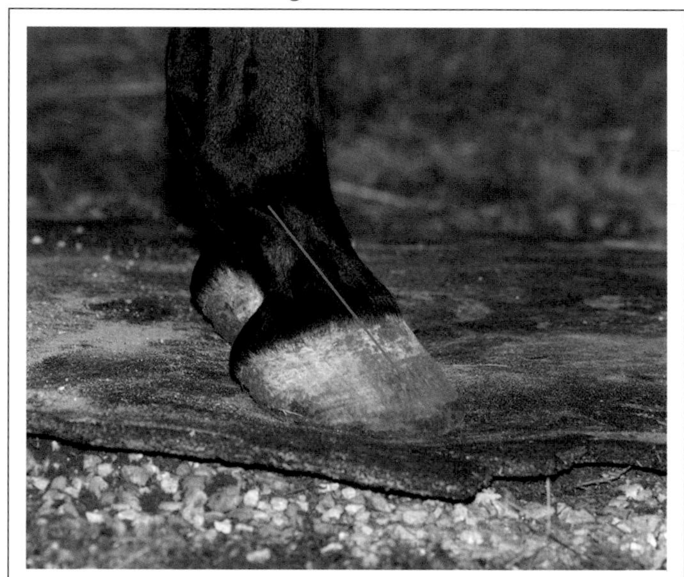

- This horse has an unbroken pastern axis.
- In other words, if you draw a line through the middle of his long pastern bone and through the hoof, running parallel to the front of the hoof wall, that line will be straight.
- A straight pastern axis like this is desirable as opposed to a broken pastern axis.

gait and is jarring to the joints and bones of the lower leg.

Ideally, the angle of the pastern matches the angle of the hoof wall. In other words, you should be able to draw a straight line along the front of the hoof wall and up the front of the pastern. The angle of this line should also roughly match the angle of the horse's shoulder blade, which is easily visible.

If you drew a straight line through the middle of the pastern to the ground, parallel to the front of the hoof wall, you would trace the line that is considered the pastern axis. Ideally, the front of the hoof wall and the pastern should be at the same angle to the ground. The pastern axis can be broken back, meaning the horse's pastern angle is steeper than the front of the hoof wall angle, or it can be broken forward, meaning the front of the hoof wall angle is steeper than the pastern angle.

Short, Upright Pastern

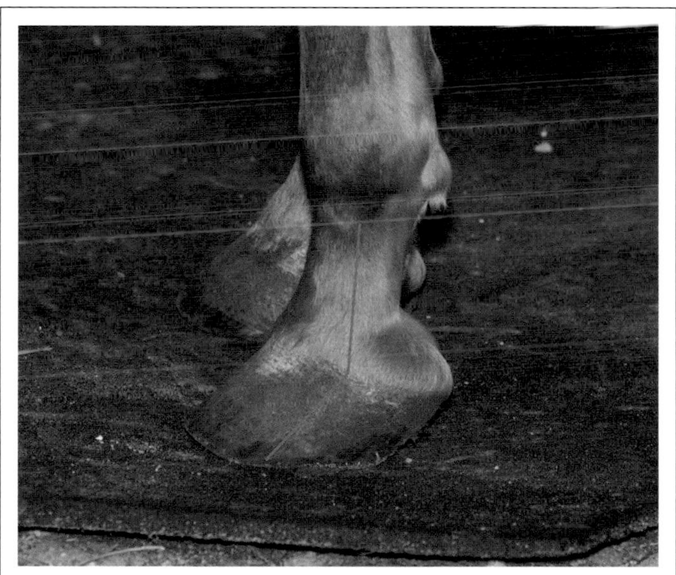

- The horse pictured here has pasterns that are too upright.
- This horse has a broken back pastern axis, with the pastern more upright than the hoof.
- A horse with short, upright pasterns will be jarring to ride, especially at the trot.
- Because short, upright pasterns are more jarring to the bones and joints, they can cause or exacerbate issues such as navicular disease or ringbone (see pages 152 and 92).

Too Long, Sloping Pastern

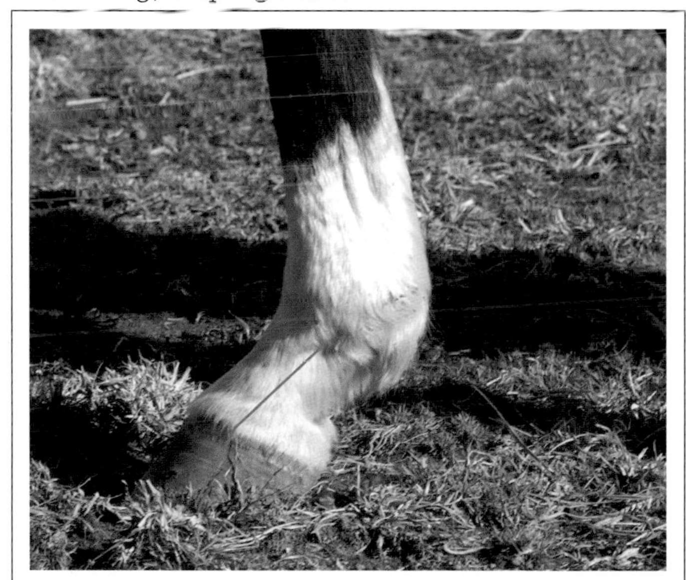

- This horse has a broken forward hoof-pastern axis, with his hoof more upright than his sloping pastern.
- Conformation faults like this can cause the fetlock to hyperextend toward the ground, placing extra strain on the tendon and ligaments supporting the lower leg during weight bearing.
- Although this horse's pasterns aren't excessively long, pasterns that slope down too much are often associated with pasterns that are too long.

HOOF SHAPE ABNORMALITIES

Long toe/low, underrun heel, and contracted heels are problems that must be addressed for soundness

Long toe/low, under run heels are extremely common and can be caused by genetics, environment, or poor trimming and shoeing practices. Long toe/low heel refers to just that: The toe of the horse is too long, and the heel is too low and may become underrun, meaning it grows toward the toe and under the horse.

Having long toe/low, underrun heels interferes with the hoof's shock-absorbing capabilities, where the heels and hoof expand. This type of hoof abnormality also places extra stress on the back portion of the foot and can cause a host of problems, including navicular issues, hoof cracks, interference issues, coffin joint synovitis, heel pain, and bruising.

A Well Maintained Hoof

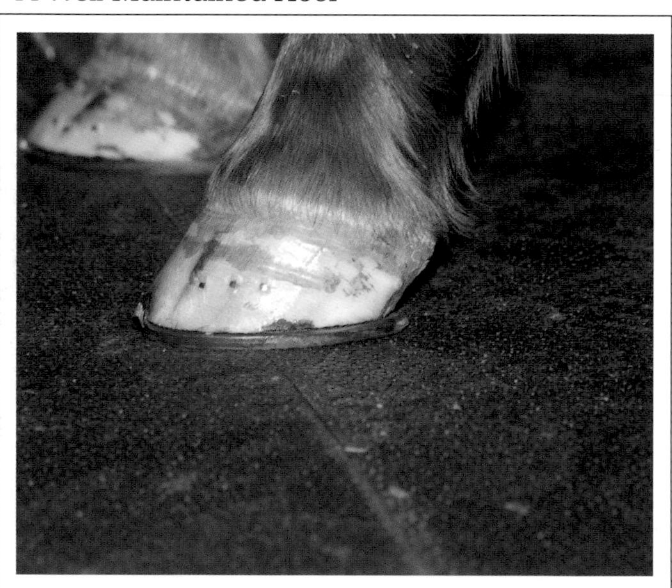

- For comparison, the photograph above shows a well-maintained hoof.
- Notice that the horse's heels are not contracted, nor are they crushed or under run. Instead you see adequate heel area without deformity.
- Regular attention from a trained farrier or barefoot practitioner is key for preventing hoof problems and soundness issues.
- In a balanced hoof, the heels will expand slightly during weight bearing, which is an important role.

Long Toe Low Heel

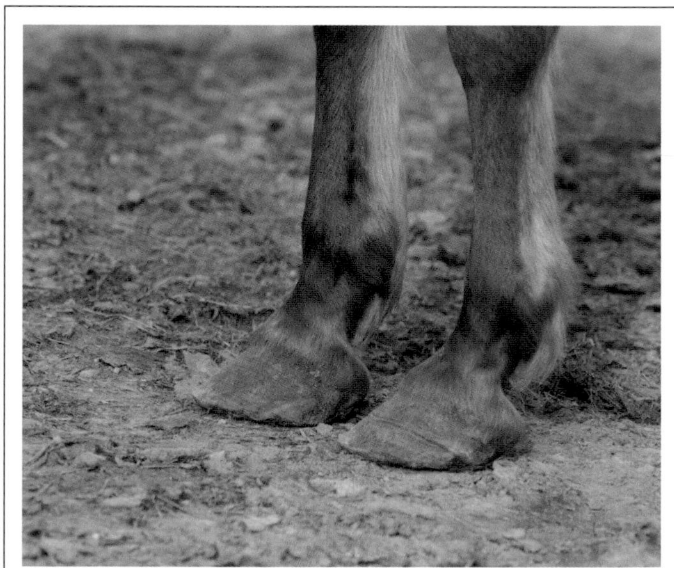

- Irregular or poor hoof care can lead to long toe/low heel; the horse's hooves in the photo above are in need of some maintenance.
- A horse with long toe/low heel has a broken-back pastern axis, meaning the horse's pastern is more upright than the front of the sloping hoof wall.
- When hooves go amiss, x-rays may be helpful for your farrier to see exactly what's going on inside the hoof.

Long toe/low, underrun heels can happen to any horse, but they're especially common in Thoroughbreds and horses with already troubled pastern angles, such as horses with too long and sloping pasterns. In addition, if a horse has a mild grade of club foot on one foot, then the other hoof may have an increased tendency for developing a long toe/low, under run heel.

Another heel problem is contracted heels, where the heels have contracted and become narrower, often pulling away from the ground along with the frog and sole. Contracted heels can be caused by lameness, neglect, not keeping the foot properly trimmed, or improper shoeing or trimming.

With either long toe/low, underrun heel, or contracted heels, it's imperative that you work with your farrier to resolve the problems and return the hooves to proper balance. There are many methods for correcting these issues. Use a skilled farrier you trust, and communicate with him or her regarding the treatment plan.

Contracted Heels

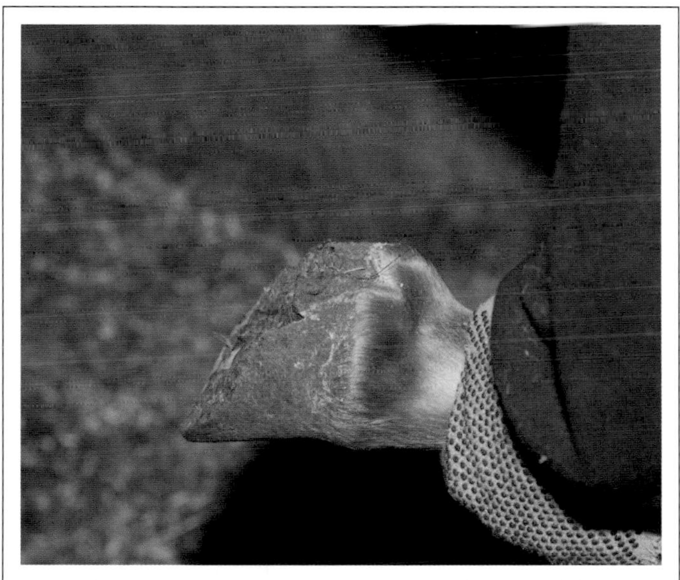

- On a horse with contracted heels, the heel bulbs are smaller and closer together than normal.
- Horses with contracted heels will not only have a narrower heel area, but the entire hoof may also be narrower than normal.
- The heel, sole, and frog all play a role in hoof expansion during the weight-bearing phase of movement, and this condition interferes with that natural mechanism.

Contracted Heels Underside

- A horse with contracted heels will often have a frog that looks shriveled up and is shrinking and pulling away from the ground.
- Horses suffering from contracted heels may also have a sole that is more concave than the soles of normal horses.
- Both horses with long toe/low heel and horses with high heels are susceptible to developing contracted heels.

MISMATCHED AND CLUB FEET

Club feet, sheared heels, and mismatched feet need careful attention

Horse owners should be familiar with several additional hoof growth abnormalities, including club feet, uneven or sheared heels, and mismatched feet. Club feet are a flexural (flexion) deformity usually involving the deep digital flexor tendon. Pain and the inability to bear weight in a normal manner can cause the hoof to deform. Club feet are most often seen in foals as either a congenital problem or a growth problem. However, older horses can also develop club feet from an injury. Club feet are very upright, with steep angles and tall heels.

Uneven or sheared heels involve the two bulbs of the heel on a hoof not matching up. When you pick your horse's hoof up, imagine a line down the middle, from the middle

A Healthy Hoof

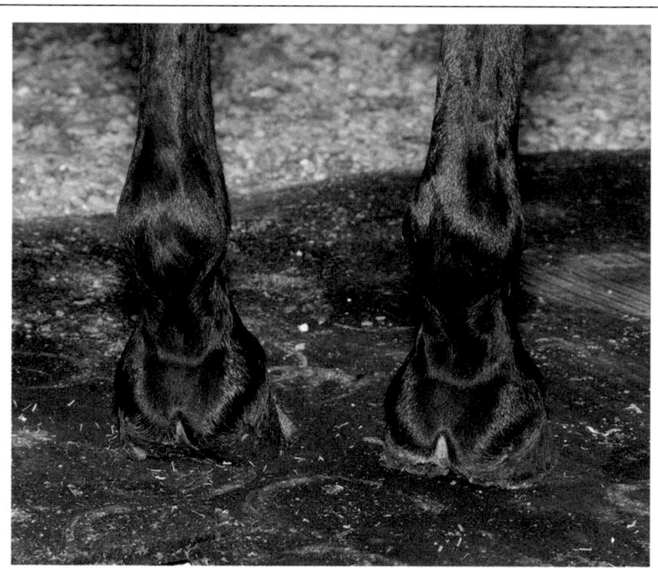

- A healthy and balanced hoof matches its pair on the opposite side and also is a mirror image of itself if you draw a line down the middle from heel to toe.
- The coronary band/coronet on a healthy foot is straight in front and slopes evenly down on each side toward the heels.
- The two heel bulbs on a healthy hoof are equal in size and height.

Club Foot

- Horses with club feet have tall, upright hooves with extra heel height.
- The veterinarian will need to work with the farrier to help correct a club foot, and x-rays may be recommended.
- Club feet can result when a foal has contracted tendons.
- Surgery to release the check ligament or one of the tendons may be necessary to help treat a club foot.

of the frog to the toe. The two sides should be symmetrical and match. Anything that causes them not to match is a problem. With sheared heels, the heels and heel bulbs do not match—there is more hoof wall on one side of the heel than the other, and one heel bulb is higher than the other. This problem can be caused by uneven weight bearing due to pain or injury to some part of the leg or hoof, or by poor trimming and shoeing practices.

Another problem is mismatched hooves, where the horse's two front feet don't match or his two hind feet don't match. For example, a horse may have a slight tendency toward a very mild grade club foot on one side, and the other side may then develop a long toe/low, under run heel shape. Mismatched feet can be genetic or created by poor trimming, or they can be caused by pain or injury to one leg or hoof, making the horse bear weight differently on each side. The goal, of course, is to correct any injury or pain and restore balance to both feet.

Sheared Heels

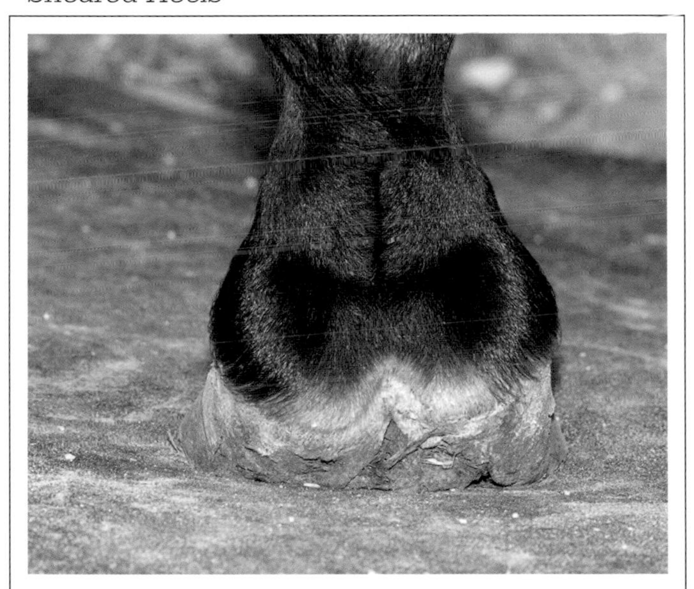

- During the weight-bearing phase of movement, a horse's heels expand.
- When a horse has sheared or uneven heels, and one heel is higher than the other, normal expansion is impaired, and one heel will bear more weight than the other.
- Careful attention by a trained and trusted farrier is needed to help correct sheared or uneven heels and return the hooves to balance.

Mismatched Feet

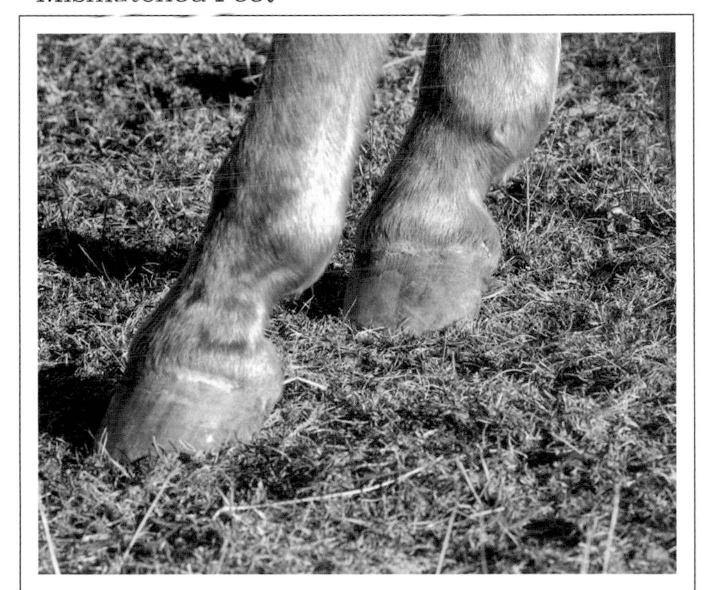

- Mismatched feet are most common in the front. This is not surprising, considering the horse's front legs bear more weight than the hind legs and are more often injured.
- To address mismatched feet, the owner, farrier, and veterinarian first must determine if pain is causing the horse to bear weight unevenly.
- After any pain or injury is addressed, the farrier can work toward balancing the feet so that the horse can bear weight and travel evenly.

BONES OF THE LEG

It's important to be familiar with the bones of the front and hind legs

There are numerous leg bones in the horse to be familiar with. The bones below the knee and hock are the same, whether you're talking about the hind leg or the front leg. Above the short pastern bone sits the long pastern bone (also known as the first phalanx or proximal phalanx or long pastern bone). Sticking out above and to the back of the long pastern bone—in the fetlock—are the proximal sesamoid bones.

Next is the cannon bone (or third metacarpal in front, third metatarsal in back). This long, straight bone is the main bone between the pastern and the knee or hock. Toward the upper back part of the cannon bone on either side are the smaller splint bones (also known as second and fourth metacarpal in front and second and fourth metatarsal in back). The splint bones do not go all the way down the

Ulna, Radius, and Carpus

- You can see the ulna, which is the upper leg bone of the elbow ("point of the elbow") in front of the girth area, in the photograph above.
- The graphics also show the long front "forearm" bone called the radius, and below it the knee or carpus.
- The carpus is made up of seven small bones. (Note: Some horses have six or eight small bones.)
- The small bones in the carpus are stacked and arranged in two rows.

Cannon and Splint Bones

- The cannon bone sits relatively straight up and down with one splint bone along each upper back side of the cannon bone.
- The splint bones are thicker on top and taper off in the lower half, before the cannon bone ends at the fetlock.
- The splint bones that are on the inner side are known as the inner or medial splint bones.
- The splint bones that are on the outer side are called the outer or lateral splint bones.

length of the cannon bone but taper off in the lower half.

Above the cannon bone in the front legs is the knee or carpus—a complex set of small bones that fit together like puzzle pieces. Between the carpus and the shoulder area comes the long, straight "forearm" bone known as the radius. At the upper back tip of the radius is the ulna, a small bone pointing toward the horse's ribs, which forms the elbow.

On the hind legs above the cannon bone is the hock or tarsus, which, like the knee, is made up of a number of smaller bones. The long tibia bone above the hock does not sit straight but juts forward where it meets the femur above.

Pastern Bone and Proximal Sesamoid

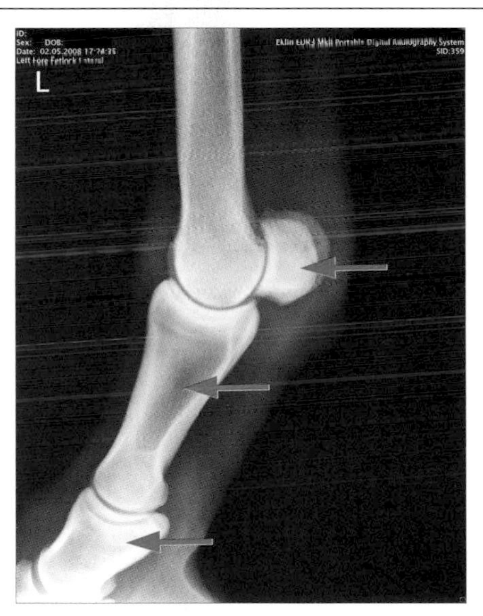

- The long pastern bone is so called because it is indeed longer than the short pastern bone below it.
- The long pastern bone sits at an angle because it hinges the fetlock downward to help absorb concussion during weight bearing.
- The fetlock is the joint where the long pastern and cannon bone meet.
- At the back of the fetlock are the proximal sesamoid bones.

Stifle, Tibia, and Tarsus

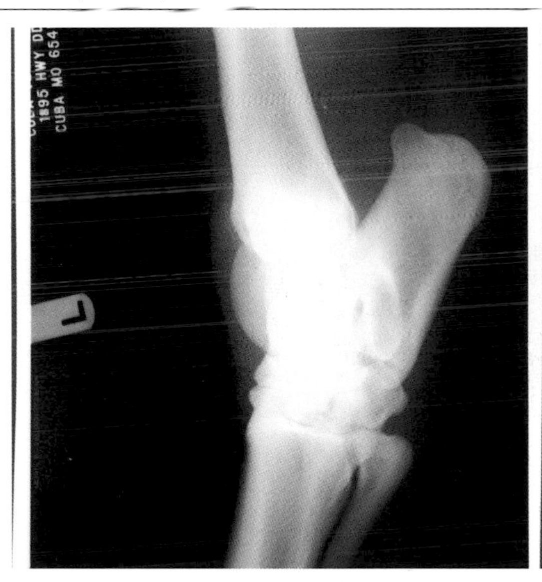

- The hock or tarsus (shown above) is made up of six small bones. Like the knee, the hock suffers a great deal of wear and tear during athletic pursuits.
- The larger bone that sticks up and out at the back of the hock is called the calcaneus.
- The tibia is the long, slanted bone above the hock.
- Above the tibia is the stifle joint.

JOINTS

Joint problems are common, so it's important to understand how joints work

Bone grinding on bone would not work well for us or for horses, which is why horses have joints. It's important to understand joints, as arthritis, which virtually all horses eventually get at some point, affects the joints. There are three types of joints, but the main leg joints are synovial joints, which are the focus for this section.

A joint is formed where two bones that need to move independently meet. The ends of the bones are coated with smooth articular cartilage. The joint is stabilized with a fibrous joint capsule that attaches to bones as well as ligaments. Most joints have collateral ligaments along the sides as part of their stability. The joint capsule also holds fluids in

Carpal Joints

- The equine knee or carpus on the forelimbs is comprised of seven small bones and three notable joints. (Note: Some horses have six or eight small bones.)
- The radiocarpal joint is between the radius ("forearm bone") and top carpal bones.
- The intercarpal joint is between the two rows of small bones within the knee.
- The carpometacarpal joint is between the lower carpal bones and the cannon and splint bones below.

Pastern and Fetlock Joints

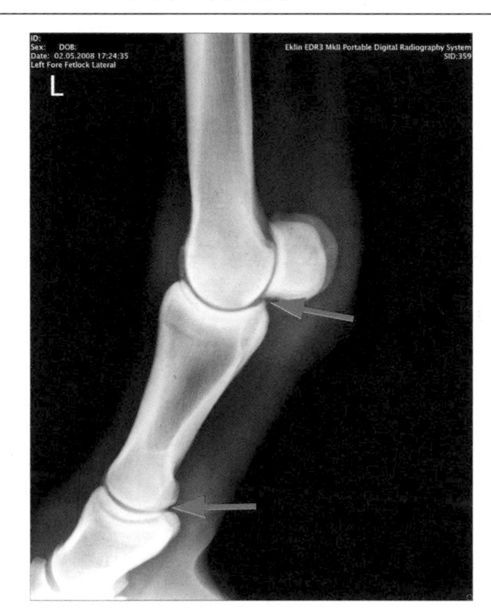

- The upper arrow above shows the fetlock joint, which is where the cannon bone meets the long pastern bone.
- The fetlock joint flexes considerably during weight bearing and is therefore subject to a great deal of wear and tear.
- The lower arrow shows the pastern joint, which is where the long and short pastern bones meet.
- The pastern joint is a lower motion joint than the fetlock.

and is lined with a synovial membrane. The synovial membrane secretes synovial fluid. Hyaluronic acid and lubricin are part of the synovial fluid makeup and serve as lubricants. The synovial fluid also provides nutrients and removes waste from the articular cartilage.

Joints are designed so that the bones can move smoothly and without friction. However, joints that must withstand a great deal of motion and concussion often experience breakdown, which we'll address in later chapters.

ZOOM

Cartilage is a form of dense connective tissue that does not contain blood vessels or nerves. There are different types of cartilage in the body. Articular cartilage is the type found in joints and helps buffer the bones from concussion. In healthy joints, new cartilage tissue is constantly produced to replace the old.

Coffin Joint

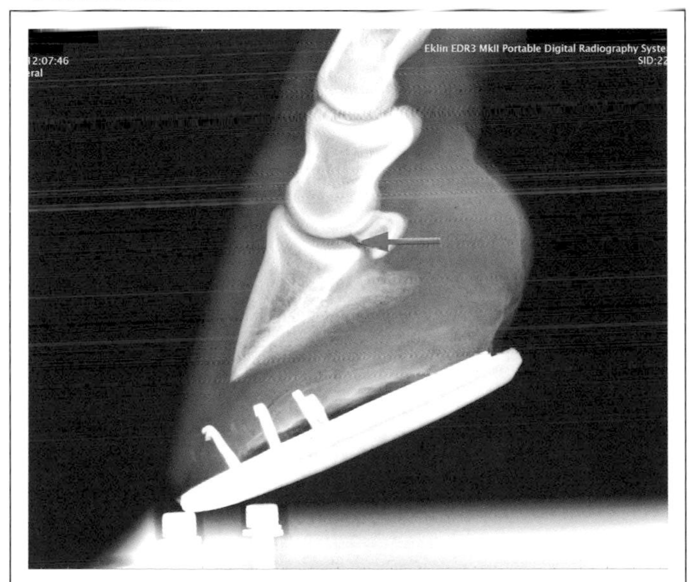

- This arrow shows the coffin joint where the bottom of the short pastern, the coffin bone, and the navicular bones meet.
- The coffin joint is the only joint within the hoof.
- The coffin joint is therefore the most distal joint.
- *Distal* is a term used to mean farthest from the body, and when it comes to horse legs, that means closest to the ground!

Hock and Stifle Joints

- The upper arrow shows the stifle joint.
- The stifle joint is located in the flank area between the long, slanted tibia bone and the femur.
- The lower arrow shows the hock.
- Like the knee, the hock is comprised of a number of bones, and there are four joints in the hock.

LIGAMENTS OF THE LEG

Ligaments hold bones together and keep them from hyperextending

Ligaments are bands of connective tissue and are both flexible and strong. Ligaments connect one bone to another, or connect bone to other tissues like cartilage. There are different types of ligaments in the horse's leg. This is because some ligaments' main job is to hold things together, while other ligaments must allow bones to come apart but not so far that they overextend.

The suspensory ligament is the most commonly injured ligament. It is part of the suspensory apparatus that keeps the fetlock joint from hyperextending and touching the ground during weight bearing. The suspensory ligament starts below the knee and runs down the back of the leg. In the lower half of the cannon bone, it splits into two—the medial and lateral ligaments—each inserting into the medial and lateral

Check and Suspensory Ligaments (Front)

- The lines in this photo show the suspensory ligament as it travels from behind the cannon bone to the proximal sesamoid bones.
- The arrow shows the inferior check ligament where it runs from below the knee toward the back middle part of the cannon bone.
- The inferior check ligament, or accessory ligament of the deep digital flexor tendon, helps keep the deep digital flexor tendon from overextending during weight bearing and helps carry its load.

Check and Suspensory Ligaments (Hind)

- Because the hock or tarsus is made up of several bones, there are a number of ligaments in the hock area not shown here.
- The lower hind leg ligaments are virtually the same as the front leg with the inferior check ligament and the vital parts of the suspensory apparatus: the suspensory ligament and the distal sesamoidean ligaments, which are so important to supporting the fetlock and avoiding hyperextension when the fetlock lowers during weight bearing.

proximal sesamoid bones (the little bones at the back of the fetlock). From there, the suspensory ligament splits into even more branches illustrated in the photos below.

The inferior check ligament, or accessory ligament of the deep digital flexor tendon, is another commonly injured ligament in the horse's front legs. The inferior check ligament starts at the back of the knee and travels under the deep digital flexor tendon until it joins it near the middle back part of the cannon bone.

ZOOM

There are a number of terms you'll need to know when talking horse legs: *Proximal* means closer to the body, and *distal* means closer to the ground. *Lateral* means toward the outside, and *medial* means toward the middle (of the body). *Dorsal* means the front of the legs, and *palmar* (forelimb) and *plantar* (hindlimb) refer to the back sides of a horse's legs.

Suspensory and Sesamoidean Ligaments

- After the suspensory ligament reaches the proximal sesamoidean bones, branches of it then travel forward to the front of the pasterns and join the common digital extensor tendon as shown by the first line.
- Additional branches of the suspensory ligament travel down to the lower back part of the pasterns where they are called the distal sesamoidean ligaments as shown by the line traveling down.

Annular Ligaments and Others

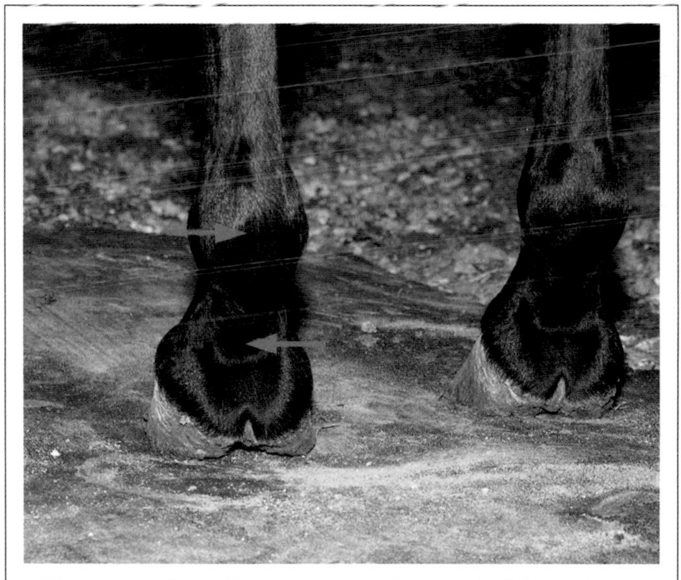

- The above photo shows two annular ligaments that wrap around the back part of the fetlock and pastern.
- One of the jobs of annular ligaments is to help keep the tendons in place.
- Keep in mind that this section has only highlighted a few of the ligaments in the leg.
- There are many additional ligaments we have not covered that play important roles and can also be injured.

TENDONS

Tendons move bones, allowing joints to flex and extend, but tendons are also vulnerable to injury

Tendons connect bone to muscle. For a horse to walk, trot, canter, or jump, he needs to move his bones, but bones don't move on their own. When a muscle contracts, tendons transfer that contraction, making the bone move.

Similar to the makeup of ligaments, tendons are long, fibrous cords of connective tissue. They are wrapped in a tendon sheath similar to a joint capsule with an inner synovial membrane where they experience friction, such as over a joint, or where they change direction.

Tendons are divided into two types: flexor and extensor. Flexor tendons allow joints to flex back, while extensor tendons allow joints to extend forward.

Tendons 1

- The arrows in the photograph above show the paths of the superficial digital flexor tendon (closest to skin) and the deep digital flexor tendon in back, as well as the digital extensor tendon in front.
- Above the knee, the superior check ligament connects the superficial digital flexor tendon to the radius ("forearm" bone).
- Below the knee, the inferior check ligament connects the deep digital flexor tendon to the back of the knee/hock.

Tendons 2

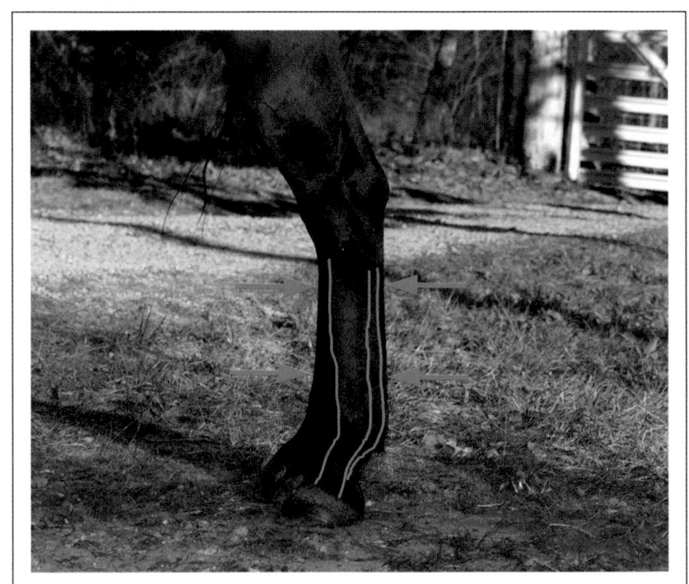

- The paths of the superficial digital flexor tendon, the deep digital flexor tendon, and the digital extensor tendon are similar in a hind leg as they are in the front leg.
- However, injuries to tendons in the front legs are more common since the front end of a horse bears more weight.
- The digital extensor tendon starts at the top front of the leg at the extensor muscle and travels down the front of the leg all the way to the coffin bone.

The two most commonly injured tendons in a horse's legs are the deep digital flexor tendon and the superficial digital flexor tendon, both of which run along the back of the legs. The superficial digital flexor tendon is the most visible or "superficial" of the two, located at the back of the lower leg, nearest the skin. The superficial digital flexor tendon starts at the flexor muscle at the top of the leg and splits when at the proximal sesamoid bones, finally ending its journey at the back of the short pastern bone. The superficial digital flexor tendon helps flex and stabilize the lower leg of the horse.

The deep digital flexor tendon starts at the top of the leg at the deep flexor muscle and travels down the back of the leg between the cannon bone and the superficial digital flexor tendon all the way to the back end of the coffin bone. The deep digital flexor tendon has the important job of helping to stabilize the lower leg during full weight bearing and to flex the lower leg as it lifts for movement.

Tendons 3

- The extensor tendon connects to the front upper tip of the coffin bone and is responsible for helping to extend the leg forward (arrow 1).
- The deep digital flexor tendon connects to the lower back part of the coffin bone (arrow 2).
- The superficial digital flexor tendon connects to the back of the short pastern bone (arrow 3).
- The digital flexor tendons help flex the leg back so the horse can bend it as needed during movement. They also help stabilize the lower leg during weight bearing.

Tendons 4

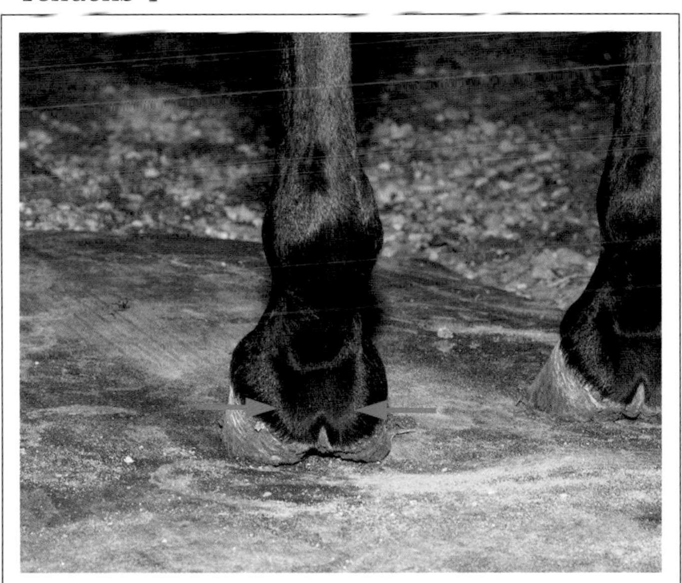

- A view of the deep digital flexor tendon from the back shows its true importance.
- There is considerable width to the deep digital flexor tendon where it sweeps under the back of the leg into the hoof capsule to connect to the lower, rear portion of the coffin bone.
- The deep digital flexor tendon is responsible for helping flex all of the lower leg joints when the leg bends.

LEG MUSCLES

The muscles at the top of the horse's leg work with the tendons to create movement

There are three main types of muscle in a horse's body: cardiac muscle powering his heart; smooth muscle lining his organs and powering involuntary internal actions; and skeletal muscle taking cues from the central nervous system to move the horse's body as needed.

The muscles powering his legs are skeletal muscles. Skeletal muscles are voluntary muscles—the brain via the nervous system tells them when to contract and how much. When skeletal muscles contract, the tendons transfer that energy to the bones, causing them to move.

Muscles can be categorized as either slow twitch (type 1) or fast twitch (type 2). In general, slow twitch muscles are needed

Muscles of the Front Leg

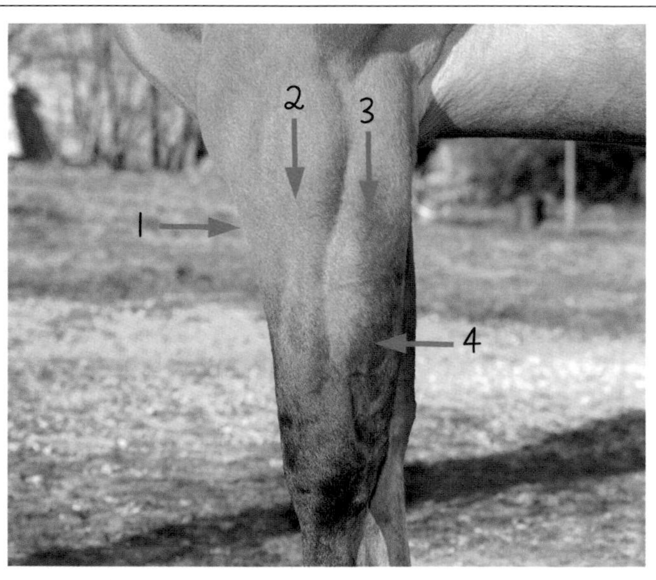

- Looking at the horse's forearm (upper part of the front leg above the knee) from the side, there are a number of muscles.
- At the front of the forearm is the radial carpal extensor muscle or extensor carpi radialis (arrow 1).
- The next muscle back is the common digital extensor muscle (arrow 2).
- The muscle toward the back end of the forearm is the lateral ulnar muscle (arrow 3), and just behind it the deep digital flexor muscle (arrow 4).

Muscles of the Hind Leg

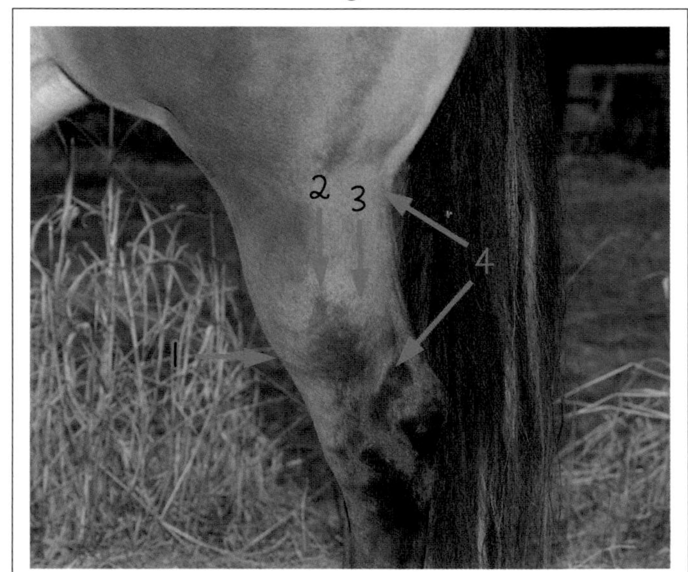

- Looking at the horse's gaskin (upper part of the hind leg) from the side, there are several muscles to note.
- At the front side of the gaskin is the long digital extensor muscle (arrow 1).
- In the middle of the gaskin is the lateral digital extensor muscle (arrow 2); and behind that the deep digital flexor muscle (arrow 3).
- At the back of the gaskin is the gastrocnemius muscle (arrow 4).

for slower work over a longer period of time, whereas fast twitch muscles are needed for quick and powerful bursts of speed. What proportion of each type a horse has is partly based on type of horse—draft horse versus Thoroughbred, for example—and partly on the horse's conditioning.

Like tendons, there are flexor muscles and extensor muscles that work in pairs, one relaxing (lengthening) while the other contracts (shortens) to bend (flex) the leg and straighten (extend) it. The muscles are in the upper portion of the horse's leg, with the tendons traveling down to the lower leg.

Muscling

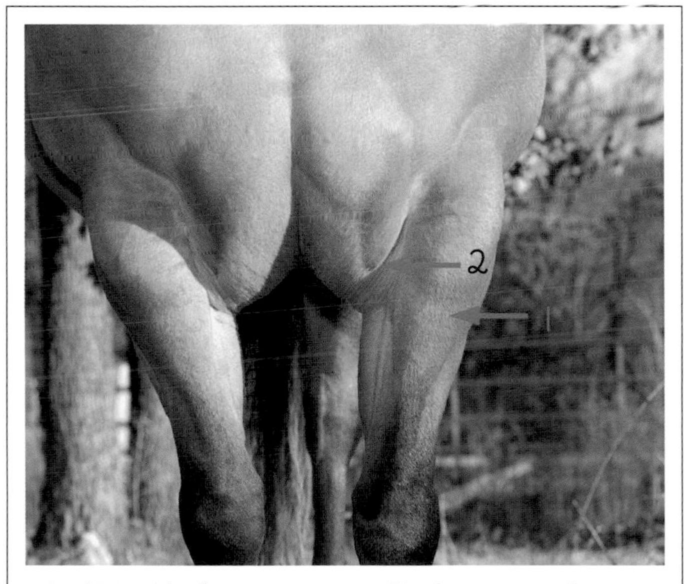

- Looking at the forearm area from the front, notice the radial carpal extensor muscle at the front of the forearm (arrow 1) and the biceps brachii muscle above it in the lower chest area (arrow 2).
- A well-muscled forearm is desirable.
- The forearm muscles should taper down to the knee, not end abruptly.

Skeletal muscles contain muscle bundles, which in turn are made up of muscle fibers. The muscle fibers themselves have bundles of myofibrils.

Muscles need fuel to work. Glucose, which comes from carbohydrates, is a main muscle fuel. Muscles use oxygen carried through the bloodstream to help them convert or break glucose down. One of the things muscles make out of glucose is lactic acid. Formerly thought to be a "bad" by-product, lactic acid is now understood to be a fuel, allowing mammals harder and longer exercise.

Extensor and Flexor Muscles

- Muscles, like tendons, can't push; they can only pull, which is why they have opposing pairs—extensor muscles and tendons and flexor muscles and tendons.
- The flexor muscles and tendons work to flex the joints, such as when the hoof is lifted during movement (as seen in the photograph above).
- The extensor muscles and tendons then work to extend or straighten the leg back out.

GOOD CONFORMATION

Horses with excellent conformation fare better and have fewer lameness issues

Conformation refers to the way a horse is built—the way he is "put together" or conformed. Ideal conformation varies by breed and intended use. Certain types of horses are better built to excel at particular sports. For example, draft horses have conformation for heavy work, such as pulling wagons or plows, while many Arabians have a build better suited to pursuits such as endurance riding. Within a breed there are also variations—such as Quarter Horses bred for cattle work as opposed to hunter under saddle.

This makes defining proper conformation difficult. However, conformation can be discussed in terms of function. Severe deviations from what's generally considered good

Good Front Leg Conformation

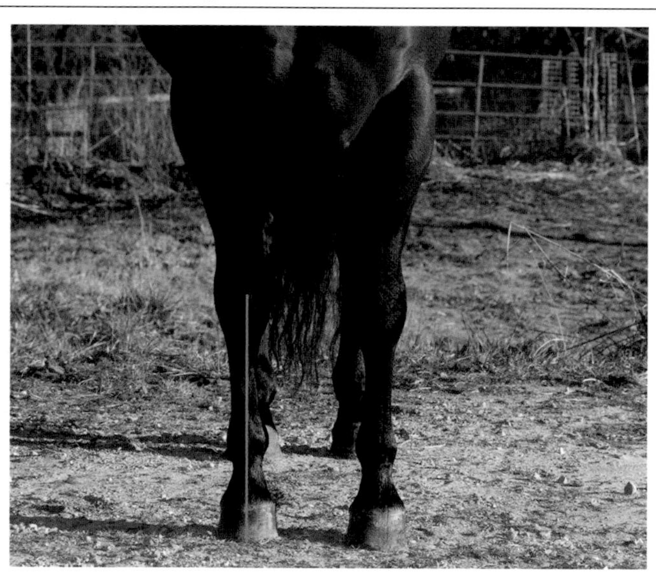

- To judge front leg conformation from the front, imagine a straight line starting at approximately the point of each shoulder and traveling down.
- The line should pass through the middle of the knee and travel straight through the middle of the cannon and pastern bones, and straight through the middle of the hoof.
- When judging conformation, it's easier if the horse is standing on flat ground, as bumps and inclines will change the angles.

Good Front Leg Conformation (Side View)

- To evaluate front leg conformation from the side, visualize a straight line as shown in the photograph above.
- The line should run through the forearm, straight through the through the middle of the knee, and straight through the mid part of the lower leg and fetlock, ending just behind the hoof.
- In halter classes, horses are judged on their conformation. Most horses are asked to stand squarely, with all four legs evenly placed underneath them for easy assessment.

conformation usually result in poorer function. Horses with conformational faults experience greater wear and tear on their muscles, tendons, ligaments, joints, and bones.

Although this book is focused on the horse's legs and hooves, keep in mind that poor upper body conformation also affects the way a horse moves and his long-term soundness. A horse that is "downhill," with his hind end higher than his front end, experiences added weight and strain on the front legs. Just as a short, upright shoulder can make for a short, choppy gait, the makeup of the horse's croup (top of the rear end from the point of the hip to the point of the buttock) and hind end also affects his movement and stride.

In addition to assessing the angles and shapes of bone and muscle on a horse, conformation analysis also examines how different body parts compare to each other—whether the horse is balanced and whether certain body parts are too small compared to others, such as hooves that are too small compared to the horse's legs and body.

Good Hind Leg Conformation

- Hind leg conformation can be assessed from the back by imagining a line like the one in the photograph above.
- The line should pass from the point of the buttock through the middle of the hock, down the middle of the lower leg, and through the middle of the fetlock and hoof.
- Although you're comparing against an imaginary ideal, it's more obvious deviations rather than subtle ones that tend to cause problems.

Good Hind Leg Conformation (Side View)

- To judge hind leg conformation from the side, visualize a line like the one shown in the photograph above.
- The line should travel from the point of the horse's rear end, along the back part of the hock, and straight down the back part of the cannon bone, landing about three inches behind the back of the hoof.
- A mat (like the one this horse is standing on) or a flat area of cement or pavement is a good place for evaluating conformation.

FRONT LEG DEVIATIONS

These front-leg conformation faults can lead to movement or soundness problems

Some conformational faults of the front legs place added strain on the legs—tendons, ligaments, and joints—and can cause gait abnormalities, which will be discussed later.

Imagine you're looking at a horse's front legs while standing in front of him. For comparison, first visualize a horse with good conformation. The line goes straight from the points of his shoulders down through the middle of his knee, the middle of his cannon bone, fetlock, and through the middle of his hoof. Now look at possible faults. If the horse's toes point out, he's splayfooted (toed out). If his toes point in, he's pigeon toed (toed in). If he's bowkneed, his knees are a bit to the outside of that imaginary line (think bowlegged), and knock-

Toed Out

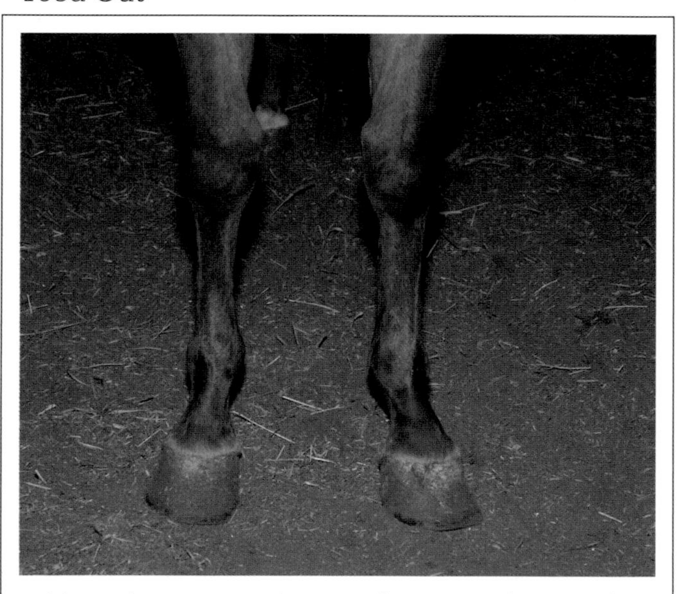

- A horse that toes out or is splayfooted like the horse in the photograph above has the toes of his hooves pointing out to the sides rather than straight ahead.
- A horse that toes out may also be base wide (hooves too far apart compared to chest) or base narrow (hooves too close together compared to chest). Both of these faults can cause interference (hoof hitting opposite leg).
- Base wide places added stress on the inside of the limb, while base narrow places added stress on the outside.

Toed In

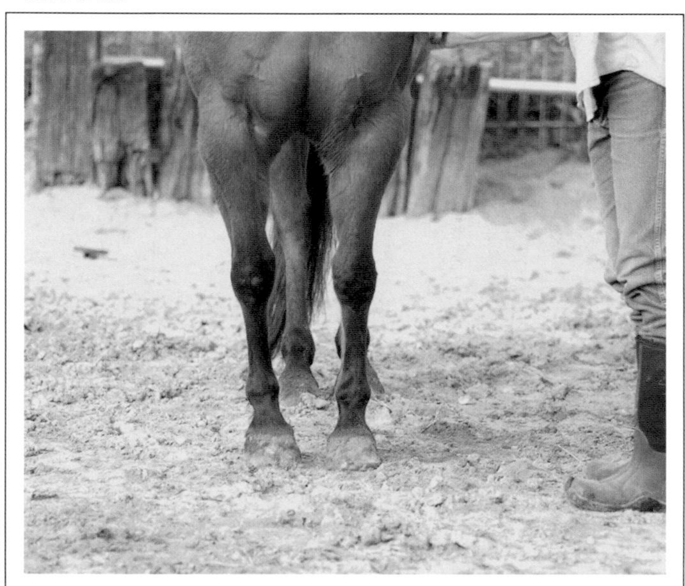

- A horse that toes in or is pigeon-toed has his toes pointing slightly inward toward each other.
- A horse that toes in may be base narrow like this horse but is less likely to be base wide.
- Toeing in is relatively common and places added strain on the fetlock and pastern joints.
- Although horses with conformational defects can be athletic, they are often more susceptible to certain injuries and strains.

knees or knee-narrow is the opposite problem. Bench-knees or offset knees means that the area from the knee down is a bit to the outside of that line; the cannon bone is offset.

Now look at the horse's front legs from the side. Again, draw an imaginary line as it should be for a horse with correct conformation—through the forearm, straight through the through the middle of the knee, and straight through the mid part of the lower leg and fetlock, ending up just behind the hoof. If the horse is over at the knees or buckkneed, his knee tips forward, as if he may buckle forward. If the horse is back at the knee or calf-kneed, his knee tips back and won't touch that imaginary line at all. A horse can also be camped under or standing under, meaning the entire legs from the elbow down are too far back, or he can be camped out, meaning the entire legs from the elbow down are too far forward.

Sometimes these conformation faults are seen in combination. Each fault causes added stress to different parts of the leg and interferes with movement in a different way.

Bowkneed/Base Narrow

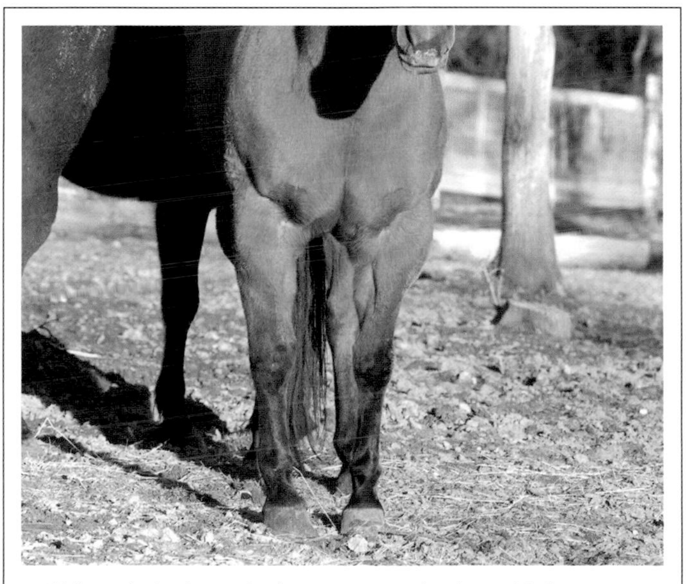

- Although the horse in the photograph above isn't standing on level ground, it's still easy to see his conformational issues.
- He is base narrow (possibly bowkneed), with his hooves set close together and his knees farther apart, rather than having his knees directly above his hooves.
- His toes also point toward each other, which is called pigeon-toed or toeing in.
- This horse appears to be more pigeon-toed on his right leg, but it may be the way he's standing or the angle of the ground.

Buckkneed

- A horse that's buckkneed (also called buckedkneed or "over at the knees") looks like his knees are too far forward when he stands straight and you view him from the side.
- A horse who is buckkneed may have his lower leg from the knee down behind the imaginary line when he's standing straight.
- Horses with buckkneed conformation place extra stress on the lower leg, including the front of the fetlock.

HIND LEG DEVIATIONS (PART ONE)

There are a number of hind leg conformational faults easily visible from the rear

Imagine you're standing behind a horse facing his hind end. First, draw that imaginary line as if you're looking at a horse with straight hind legs. The line starts at the points of the rear end on either side of the tail and passes through the middle of the hock, down the middle of the lower leg, and through the middle of the fetlock and hoof. Next, look at how that line changes when a horse has hind-end conformational faults.

Just as with the front feet, the horse may be splayfooted/toed out in back, with his toes pointing out, or be pigeon-toed/toed in, with his toes pointing in. However, keep in mind that hind legs are not designed to be 100 percent straight, so the toes pointing out slightly is normal.

Base Narrow

- The horse in the photograph above is clearly base narrow, with his hooves and lower legs much closer together than the centerline of his thighs.
- Horses that are base narrow may also be referred to as "standing too close."
- Because the weight is not evenly distributed and the legs are not underneath the horse evenly, more weight is placed on the outsides of the legs when a horse is base narrow.

Splayfooted/Base Wide

- The horse in the photograph above is severely splayfooted, which means his toes point out much more than they should.
- He is also base wide, with his hooves farther apart than the centerline of his thighs.
- In addition, this horse appears to be slightly cow-hocked, with his hocks turning out a bit and pointing toward each other.
- It's important to assess conformation before purchase and breeding the horse, since conformation affects soundness and movement.

As in the front, a horse can also be base narrow or base wide in his hind end. If a horse is base narrow, it means that centerline through his hooves is inward of the centerline at his thighs. This places a great deal of strain on the outside of his legs and hooves. A horse that's base narrow may also be bowlegged/bandylegged, with his hocks wide apart.

If a horse is base wide, that centerline in his hooves will be outside the centerline of his thighs. A horse that's base wide may also be cowhocked, which means his hocks are too close together, ends pointing toward one another, toes pointing out. When a horse is cowhocked, it places a great deal of strain on his hock joints.

These faults may occur alone, or a horse may have several faults, and the severity of the deviation varies. For example, a horse that is only slightly cowhocked experiences less strain placed on his legs and less effect on his movement than a horse that's severely cowhocked.

Cowhocked

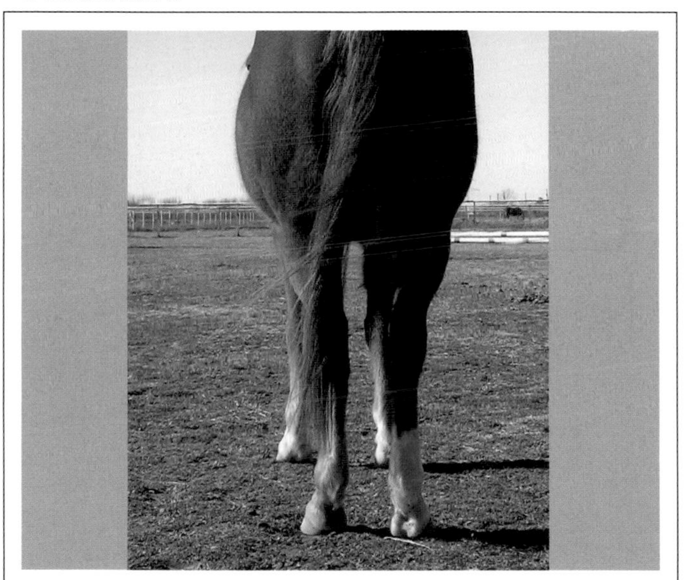

- If a horse's hocks point inward toward each other and are too close together, he's said to be cowhocked.
- A severely cowhocked horse often has trouble working off his hindquarters and performing movements such as quick turns and sliding stops.
- The opposite problem is bowlegged or bandy legged, when the hocks are wider apart than the hooves.
- On a bandylegged horse, the hocks may point out and the toes in.

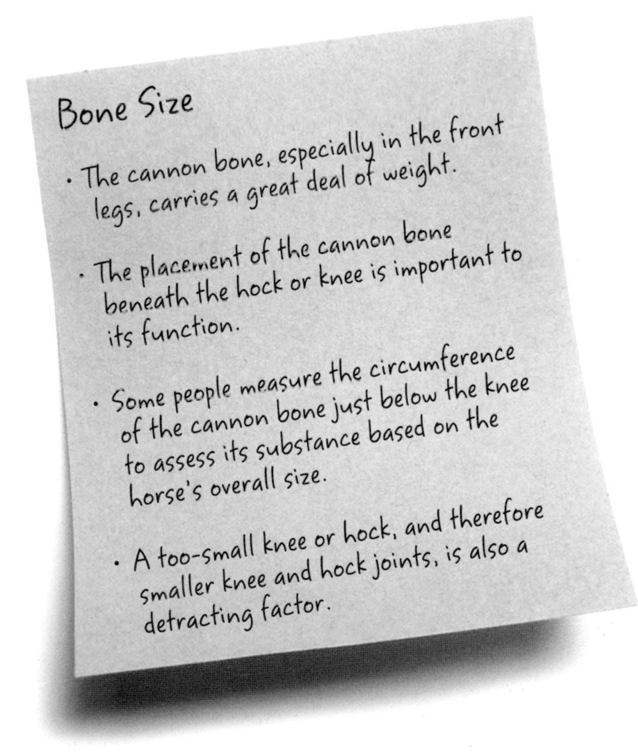

HIND LEG DEVIATIONS (PART TWO)

How hind legs are "set" will affect a horse's soundness and performance

Viewing the hind legs from the side, there can be a number of problems with how the leg is set. For reference, imagine a horse with correct hind leg conformation. From the side, visualize a straight line running from the point of the buttocks, along the back side of the cannon bone and to the ground about 3 inches behind the hoof. If, instead, the entire leg is well forward of that line, the horse is considered camped under or standing under. A horse with sickle hocks will also be standing under due to the sharp angle of his hock, which causes his lower leg to be placed well forward. Being sickle hocked can reduce a horse's stride and also places added stress on the hock.

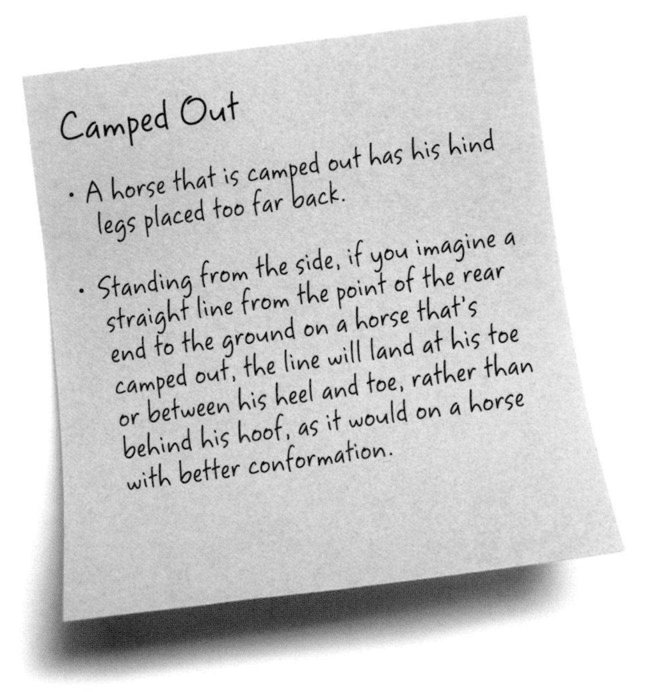

Sickle Hocked/Camped Under

- Standing from the side, if you imagine a straight line from the point of the horse's buttocks to the ground, a camped under horse's legs would be well in front of that line—too far underneath the horse.
- When a horse is sickle hocked, like this horse, he will only be camped under from the hock down because the angle of the hock is sending his lower leg forward.

If part of a horse's hock and most of his lower leg fall behind that imaginary line, he is camped out. Horses that are camped out often also have upright pasterns in the hind end. Horses that are postlegged or straight behind may also have upright pasterns. Postlegged horses have hind legs that are too straight, meaning there's not enough angulation to the tibia bone, and the hock is also very straight. Being postlegged can place additional concussion on the hock.

ZOOM

When you hear talk of high hocks versus low hocks, it's referring to where the hock is in relation to the ground (and therefore the length of the cannon bone). Generally, low hocks are more desirable because they make for smoother hock action and enable the horse to get the hocks under his body more easily.

Postlegged

- When looking at a horse's legs, certain bones are meant to have an ideal or near ideal angle. If that angle is too severe or instead there is not enough angle, it can be a defect.
- This horse's tibia ("thigh") and hock are very straight, referred to as postlegged, which can place added wear and tear on the hock.
- He also has a broken forward hoof-pastern axis, with upright hooves and very sloping pasterns.

High Hocks Versus Low Hocks

- What's considered ideal in terms of the length and angle of particular bones in a horse's legs depends somewhat on the horse's breed and intended use.
- A low hock is especially desirable for Quarter Horses and other stock breeds used for reining—the lower hock providing the leverage and power needed for the stops and turns.
- A higher hock means snappier hock action, which can be desirable for certain gaited breeds.

CONFORMATION AND MOVEMENT

Conformational faults can cause winging, plaiting, paddling, and other gait abnormalities

Conformation affects the way a horse moves, from the length of his stride to how he lifts his legs and places his feet. A horse with good conformation usually tracks straight, meaning that each hoof stays in its own "lane" or track as it's lifted forward and then placed down. These two lanes run parallel, and if you see the hoof marks left in the sand, the toes will point forward. However, certain conformational faults can lead to gait abnormalities. The severity of the gait abnormality depends on the how much the horse's conformation deviates and what conformation faults are present.

Horses that are splayfooted/toe out often wing to the inside. This means that as they travel—pick up their feet and

Straight Action

- A horse with good conformation will track fairly straight. That means the two left hooves will stay in their "lane," and the two right hooves will stay in theirs, each toe pointing more or less forward, without the leg swinging to the inside or outside as it moves forward.
- If you walk or trot your horse across a freshly dragged arena, you can view his action by the tracks left in the dirt.

Interference

- A horse's conformation affects his movement and the way he travels or places his feet.
- *Interference* refers to one foot hitting another during movement, and certain conformation faults make interference more likely.
- Both plaiting and winging can lead to interference injuries.
- Interference may be inconsistent and can be exacerbated by certain movements asked for during a workout, by footing, or by overexertion.

move them forward—they swing their feet in an inward arc before setting them down (toes facing outward). Horses with base wide conformation may also wing.

Plaiting means the horse moves his legs in an inward arc placing one foot directly in front of the next. Horses that are base narrow and toe out are prone to plaiting.

Horses that toe in often paddle, or dish. This means that as they pick up their feet, their feet swing outward before landing (toes facing slightly in). Horses that are base narrow may also paddle.

A horse is said to be forging or overreaching when the hind foot comes into contact with the front foot—usually as the hind foot is coming in for a landing, and the front foot has not yet moved out of the way (most commonly seen during the trot). Overreaching is partly a balance issue that can be affected by a number of things, including shoeing, footing, and the horse's conformation and training.

Plaiting or Winding

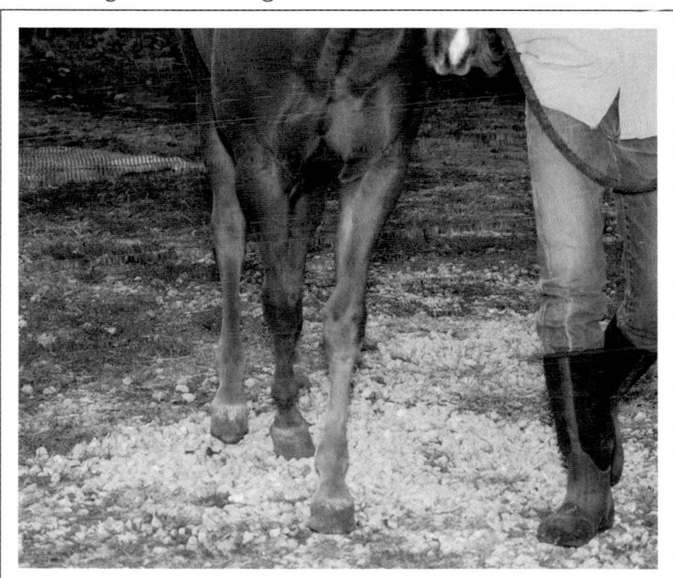

- The terms plaiting, winding, threading, and rope walking are all used to describe the action of a horse that places the lifted leg down directly in front of the supporting leg.
- The horse in the photo above shows what plaiting looks like.
- This action increases the horse's likelihood of interfering, which often leads to injury.
- Horses with an overly wide chest may be more likely to plait/wind.

Conformation and Action

- Watching a horse trot in a straight line away from you can let you assess his way of going.
- This technique is often used by halter class judges, where the judge asks each horse to trot away from him in a straight line to check for things like winging, plaiting, or paddling.
- While gaited breeds have their own desirable way of lifting and moving their legs during each gait, for non-gaited breeds, straight action is desired.

FINDING A GOOD FARRIER

Finding and keeping a good farrier is the best thing you can do for your horse's hooves

Finding a veterinarian is largely a matter of who is in your area and is accepting new clients. Unlike veterinarians, who are held to nationwide standards, farriers do not have a uniform set of standards to determine qualifications, so it's up to the horse owner to find the most qualified farrier. (Here the term *farrier* means both those that shoe and barefoot practitioners.)

There are two main ways farriers are educated: One is to attend a school. School certification and degree programs in shoeing and trimming are run by both private entities and public community colleges. Courses can last anywhere from a couple of weeks to a couple of years. One can also apprentice and work under the guidance of an experienced farrier,

Communication Is Key

- Farriers are busy people with jam-packed schedules, but it's important to keep the lines of communication open between horse owner and farrier.
- Let your farrier know any concerns you have about your horse's hooves or his soundness.
- Your farrier should be aware of the work your horse performs and on what type of footing.

Regular Shoeing

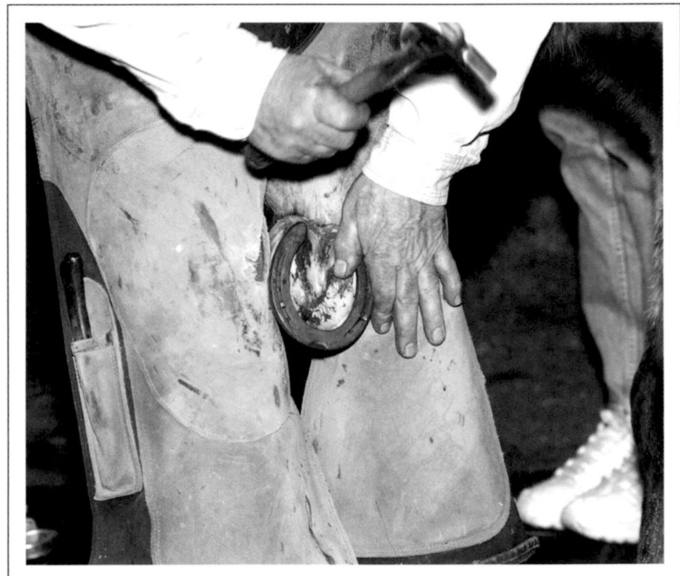

- Shoeing can be expensive, but waiting too long between shoes will cost your horse and your wallet far more in the long run.
- Most horses need new shoes every six to eight weeks.
- Depending on the type of shoe used and how much wear and tear the shoe received, occasionally a shoe can be reset after the hoof is trimmed. However, most horses require new shoes each visit.

either in place of school or in addition to school. There are also associations, such as the American Association of Farriers and the Brotherhood of Working Farriers Association, as well as barefoot groups, that offer testing and standards that result in certification.

There is no one right way for a farrier to be educated, but it is important for a farrier to be knowledgeable and capable. Most horse owners find their farrier through word of mouth. Ask local stable owners and horse trainers whom they recommend. Your veterinarian is another excellent person to ask for a recommendation. Then, once you find a candidate, don't be shy about inquiring as to his or her training and shoeing or trimming philosophy. In addition, horse owners interested in trimming their own horse's hooves should undergo training just as they would expect from a paid professional.

Regular Trimming

- Just like shod horses, barefoot horses need regular visits from the farrier to keep the hooves in balance.
- Some barefoot horses do better having their hooves trimmed more often—every four to five weeks—rather than the standard six to eight weeks.
- Your farrier can help you determine the schedule that's best for your horse's individual needs.
- Most farriers will schedule the next appointment on the spot so that you can both mark your calendars.

Keeping a Good Farrier

- Once you find a good farrier, keep him or her happy. Busy farriers can drop clients that do not treat them right.
- Have your horse caught and ready for the farrier when he or she arrives at the set appointment time.
- Have a flat, shady area for your farrier to work in.
- Above all, always pay your farrier on time! This usually means having a check or cash ready to hand over during the appointment.

SHOES OR BAREFOOT?

Deciding between shoeing your horse and maintaining him barefoot is largely an individual decision

The "barefoot movement" has been gaining momentum over the past twenty years. The general idea is that shoes are unnatural and interfere with natural hoof expansion and traction, among other things. Many barefoot proponents believe all horses should be barefoot all the time. Other horse experts believe that horses performing certain types of work, horses with certain hoof or soundness problems, or horses with conformational issues benefit from shoes. You must decide, so evaluate your horse's health and lifestyle.

If your horse currently has shoes and you decide to give barefoot a try, remember there will likely be a transition period as your horse adjusts, especially if he must work on hard

Shoes and Hard Ground

- Work over rough terrain or on hard ground has traditionally been one reason for shoeing a horse. Indeed, if hoof wear is greater than hoof growth, shoes will be necessary.
- However, barefoot horses accustomed to work on rocky or rough terrain (and with adequate hoof growth to withstand it) usually do quite well and may actually have better traction than shod horses.
- Even some police horses are now being maintained barefoot.

Barefoot and Hard Ground

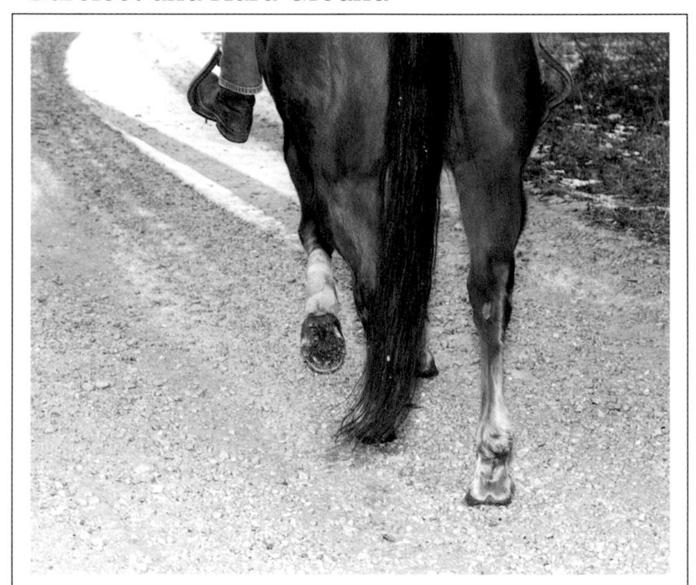

- Many horses can become accustomed to working barefoot on hard ground or on rough terrain.
- Horses that spend most of their time in a soft, fluffy stall and are then asked to work for an hour or two on hard ground may have a harder time going barefoot than horses who live in paddocks/pastures with terrain closer to that they'll experience out on the trail.
- Some owners pull their horses' shoes in winter to give the hooves a break. Others may remove shoes in the winter to begin the adjustment process.

terrain (such as on the trails). During this adjustment period, your riding time may be minimized. Some horses may never properly adjust, or their hooves will not toughen up enough for rough terrain. These horses may benefit from shoes.

People who maintain their horses barefoot often prefer to work with a farrier specializing in barefoot trims who is well-versed in maintaining barefoot horses.

ZOOM

Hoof boots are kind of like tennis shoes for your horse. Some people who maintain their horses barefoot use hoof boots on trail rides or on rough terrain. Others use hoof boots to help their horse transition from wearing shoes to going barefoot. Shod horse owners sometimes carry a hoof boot on the trail as a spare in case their horse throws a shoe during a ride.

Corrective Shoeing

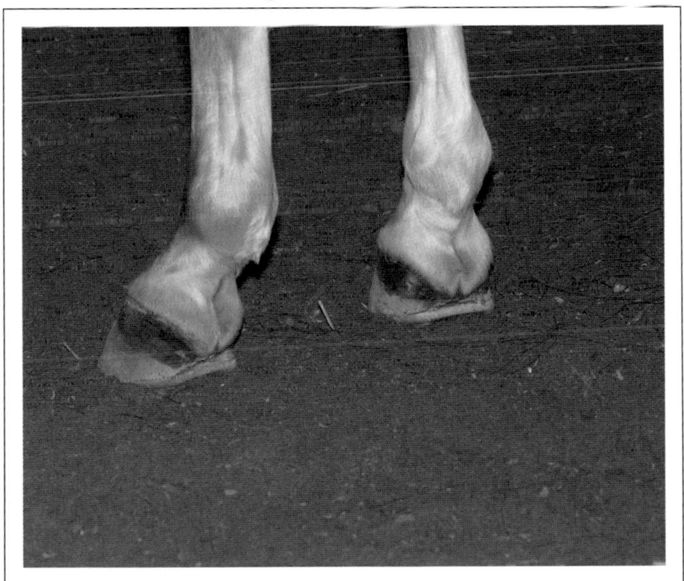

- Nowadays there are countless therapeutic and corrective shoeing options.
- Horses with certain conformation defects or lameness issues may benefit from corrective or therapeutic shoes.
- If you work with your veterinarian to handle a lameness issue, he or she may recommend pulling your horse's shoes or trying a certain type of therapeutic shoeing method.

Hoof Boots

- Several companies now manufacture hoof boots, including Easyboots and Old Mac's.
- It's imperative that a hoof boot fit correctly.
- If you order a pair of hoof boots, be sure to follow manufacturer instructions for measuring your horse's hoof before purchase.
- Each type of hoof boot goes on and adjusts differently. Adjusting the boot correctly is very important so that it doesn't wiggle or slide during movement.

BAREFOOT TRIMS

Methods and theories abound regarding the best way to trim a horse's hooves

There are myriad trimming methods and theories for horses. Some of these methods and theories are based on the study of feral horse hooves and their natural shape and wear.

One of the first to study wild horse hooves in the 1980s was Gene Ovnicek, who then developed the Natural Balance trim. Although Ovnicek also supports shoeing, many use his Natural Balance trim guidelines for barefoot horses.

A controversial barefoot advocate with a strict method of trimming and lifestyle guidelines is Dr. Hiltrud Strasser, who developed the Strasser Method. Although many agree with her recommendations that horses need to live a more natural lifestyle, the controversy arises from the rigid trimming

Four Point Trim/Natural Balance

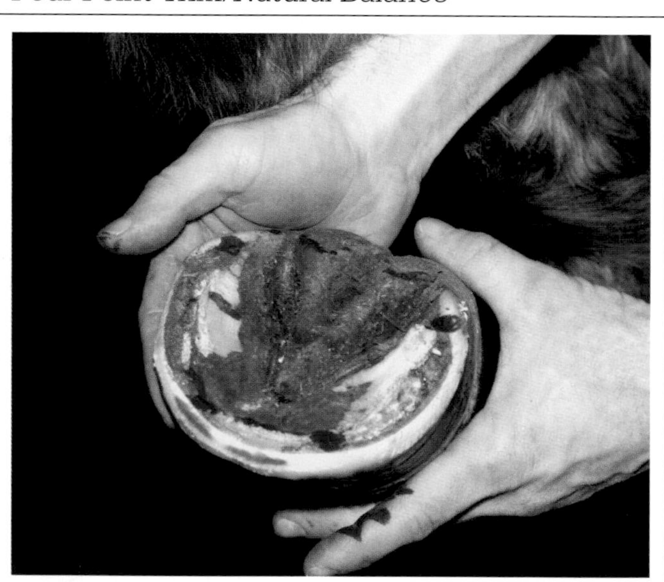

- The term *four point trim* can mean different things and is associated with more than one trimming style.
- The term has been used in association with master farrier Gene Ovnicek's trimming and shoeing method called Natural Balance.
- Ovnicek's research on wild horses showed four primary points where the hoof makes contact with the ground—one on each heel and one each at the medial and lateral (inside and outside) toe quarters.

Breakover

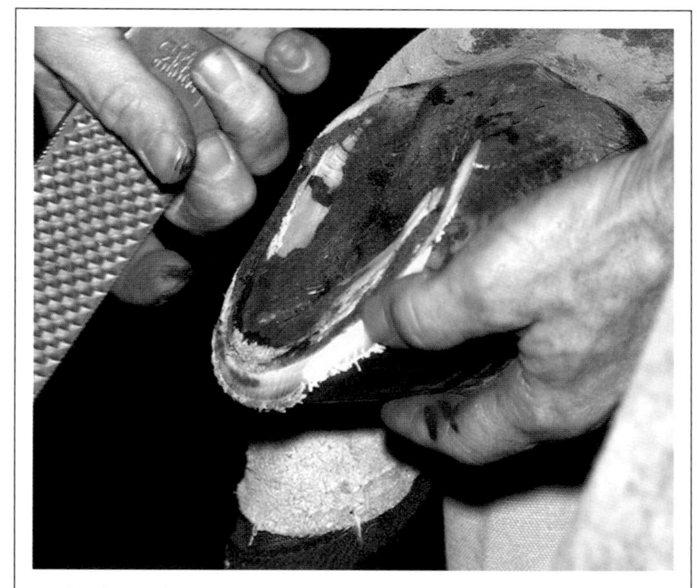

- The farrier here practices Natural Balance trimming, but farriers use different variations of the four-point trim.
- Veterinarian and farrier Ric Redden advocates four-point trimming and shoeing, including a shorter toe.
- Most trimming (and shoeing) methods take into account breakover—how and when the horse's heel and toe leave the ground during movement.
- The breakover point is generally considered to be the point where the toe leaves the ground.

guidelines and the fact that some horses are quite sore while adjusting to the Strasser trim.

There are many other notable experts, associations, and schools that have developed their own types of trims and have certification programs teaching their methods. There are also barefoot practitioners and farriers who have studied a variety of theories and adapt their trimming methods to suit the individual horse and circumstances. Most barefoot practitioners agree that a more natural lifestyle is better for horses and their hooves.

RED ● LIGHT

Although several trimming methods are mentioned here, this book does not attempt to recommend any trimming or shoeing method over another. As with the decision to shoe or not, choosing a trimming method and any accompanying lifestyle changes is an individual decision. See the Resources section of this book for a listing of Web sites that will help you in your research.

AANHCP 1

- These two photos, provided by the Association for the Advancement of Natural Horse Care Practices (AANHCP), demonstrate before and after examples of natural horse keeping and trimming methods.
- Jaime Jackson is a natural hoof care practitioner, author, and one of the founders of the AANHCP.
- In this photograph above of a hind hoof, Jackson notes the shoes have just been removed and the hoof is thin-walled, contracted, has the lateral wall broken out, a pocked sole, and virtually no concavity.

AANHCP 2

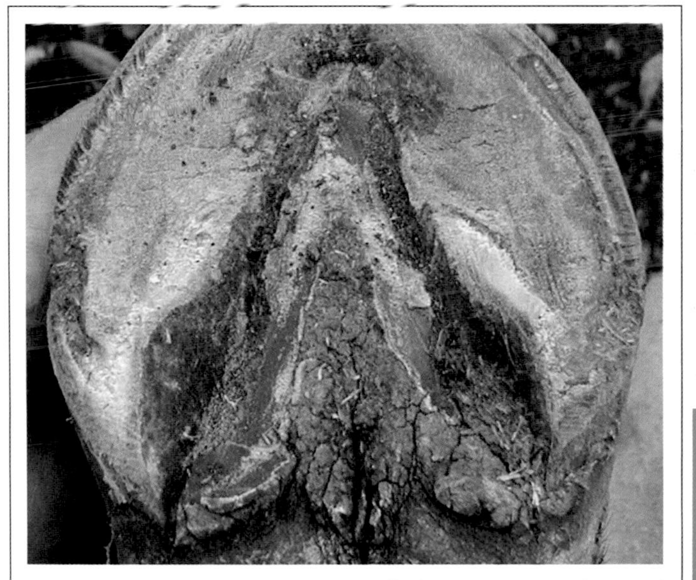

- Five months later, Jackson notes in this photo that the same hind hoof has decontracted and has a 30 percent increase in three-dimensional mass.
- Jackson, who trims on a four-week schedule, has studied wild horses and uses his research as a model for his trimming and natural horse keeping guidelines.
- The AANCHP provides Natural Hoof Care Practitioner training and certification programs, and lists Certified Practitioners on their Web site. (See the Resources section for further information.)

TYPES OF SHOES

How a shoe is made and what it's made of are important to a shod horse's performance

Proper trimming is important to shod horses, just as it is to barefoot horses. The foot must be carefully trimmed before the shoe is placed. For trimming, farriers use nippers, which are like giant fingernail clippers. They also use a rasp, which is like a giant nail file. A hoof knife may be used to pare away extra frog and sole (however, this is usually done more conservatively on a barefoot horse). Farriers then shape the horseshoe to fit the horse using a hammer and anvil (cold shoeing) or with the added use of a forge (hot shoeing).

There are several choices of material, and almost endless variations in shoe shape and style. Your farrier will help decide what shoe is best for your horse based on the type of

Steel and Aluminum Shoes

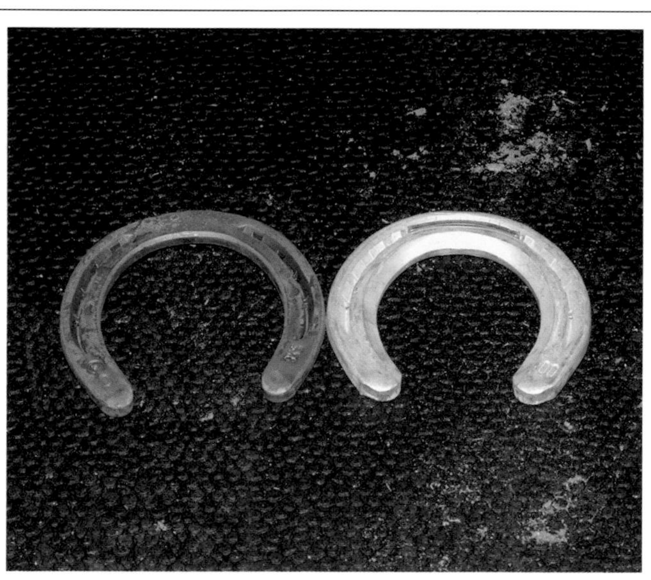

- Steel is the most common material for making horseshoes.
- Steel is strong, easy for farriers to work with, lasts long, and is the least expensive of the available shoe material options.
- Aluminum shoes are now popular for many sports horses—from dressage horses to jumpers and barrel racers—because they are lighter than steel shoes.

Shoe Clips

- If a horse is apt to throw a shoe (have it fall off) or has trouble holding nails, clips are often used.
- A single toe clip is shown here. Another type of clips are quarter clips, which are usually used in pairs—one on either side.
- Toe clips must be carefully fitted and adjusted by a skilled farrier.
- If a horse loses a shoe with clips, try to find it promptly to avoid the sharp clip being stepped on.

work he does and his individual hooves and soundness; however, it's helpful to know some of the options and their benefits and drawbacks.

Steel is the most common material for farriers to work with, and aluminum is second. Steel has the advantage of lasting longer and costing less, while aluminum is considerably lighter but can be harder for the farrier to work with. Titanium may be used on the front feet of speed horses, such as racehorses. Titanium is light like aluminum but stronger. The drawback of titanium is its price.

There are also plastic, urethane, and polyurethane nail-on shoes. Some claim to improve traction and shock absorption. However, plastic shoes are still most commonly used for therapeutic purposes, which we'll discuss in the next section along with bar shoes.

Clips can be used to help keep the shoes on. Clips are small triangles of steel either hand drawn up from the shoe when hot shoeing or factory added to a shoe. Clips are usually placed along the front of the shoes as toe clips or along the side as quarter clips.

Specialty Shoes: Reiners

- Most riding disciplines have shoes that are favored for one reason or another, and sport horses can sometimes benefit from specialty shoes.
- The sliding stop is a signature move for reiners, and reining horses are often fitted with a sliding plate on their hind feet to make the sliding stop easier.
- Sliding plates are wide, flat shoes with extended heels or "trailers."

Specialty Shoes: Barrel Racers

- Barrel racers need two things: light shoes so they can run, and a bit of grip so that they can turn.
- Because aluminum is lighter, many barrel racers opt for aluminum shoes, such as aluminum racing plates or rim shoes.
- Rim shoes feature a rim or groove running along the shoe for improved traction. On barrel racing shoes, the outside rim is usually higher.
- Polo ponies also often use rim shoes but with a higher inside rim.

THERAPEUTIC SHOES

There are a variety of shoeing options for horses that need extra support

The shoes discussed below are designed to give the horse added support or to accomplish therapeutic results that traditional shoes cannot. In addition to these basic categories of shoes, there are hundreds of specialty therapeutic shoes and thousands more custom therapeutic shoes.

The most common types of support shoes are bar shoes. Rather than ending at each heel, like traditional shoes, bar shoes go all the way around. The egg bar shoe is oval shaped and provides extra heel support. Another type of bar shoe is the heart bar, which instead of closing behind the heels follows the outside of the frog. Heart bars may be used on horses suffering from laminitis to help support the coffin

Pads

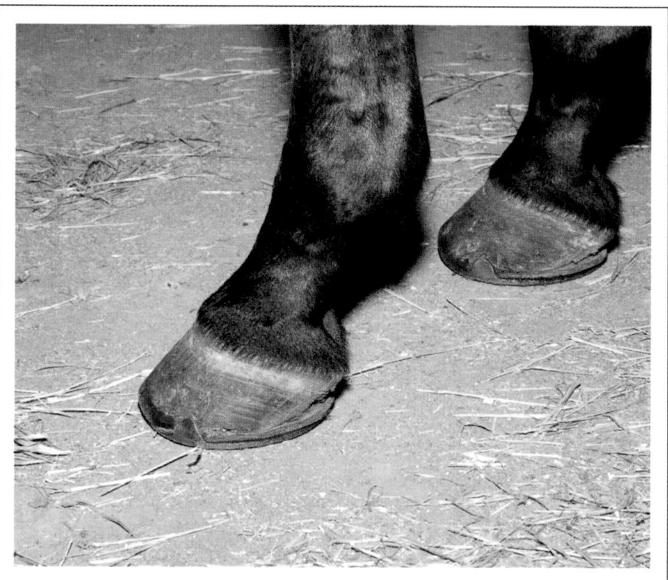

- Hoof pads may be used to protect the sole, for added support, or to reduce concussion.
- Hoof pads are usually made of plastic or another synthetic material but can be made of leather.
- Hoof pads may be packed to fill the space between the pad and the sole and frog grooves. There are a variety of packing materials.
- One downside to full pads is that they can seal in moisture and cause soggy feet.

Wedge Pads

- Pads, such as wedge pads, can alter hoof angles.
- There are also wedge shoes, where the back of the shoe is built up slightly.
- Wedge pads/shoes are used for horses suffering from navicular disease or syndrome.
- Studies show that heel wedge pads can decrease movement and strain on the coffin joint and deep digital flexor tendon, easing force on the navicular bone. However, force on the suspensory ligament is increased.

bone via the frog. A similar shoe is the full support egg bar shoe, which features a solid triangle over the frog and can be used on horses with under run heels, dropped soles, or horses who have had parts of their hoof wall removed. In addition to added support, bar shoes help prevent the horse's hoof from sinking into soft ground, as pulling the hoof back out can cause added strain.

Shoes with extended heels also add support to the heel area and may be recommended for horses suffering from under run heels. Wide web shoes are wider than traditional shoes and can be used to protect the perimeter of the sole but may decrease a horse's traction. Glue-on plastic shoes are useful for a variety of therapeutic purposes, most notably because they require no nails.

A hospital plate is a solid metal or plastic plate often bolted to the shoe to cover the entire bottom of the foot. It can be removed so that the handler can treat problems that need protection, such as puncture wounds and sole abscesses. Hospital plates may also be used for horses with dropped soles or protruding coffin bones.

Egg Bar Shoe

- Egg bar shoes encircle the hoof, providing added heel support, and are quite popular.
- Egg bar shoes may be used on horses suffering from a variety of heel problems, such as under run heels, sheared heels, or contracted heels. They can also be helpful to horses with navicular disease or syndrome.
- Bar shoes hold dirt and debris more than regular shoes, so it's important to clean the hooves regularly.

Glue-On Plastic Shoes

- There are many varieties of plastic, urethane, and polyurethane glue-on shoes available.
- Glue-on plastic shoes help horses with damaged or weak walls incapable of holding nails, and foals, whose hoof walls are not ready to take nails.
- Glue-on plastic shoes are also used on horses suffering from laminitis, because their application is far less traumatic than nails to horses that are experiencing foot pain.

SHOEING FOR TRACTION

Traction can be needed on slippery surfaces, but always add traction devices with caution

When a shod horse works on slippery surfaces such as ice, wet grass, or mud, he may need extra traction added to his shoes to keep him from constantly slipping. Traction can come in many forms, including studs/caulks, toe grabs, rim shoes, and borium. However, too much traction can have a jarring affect on a horse's movement and cause injury. It's important to use traction cautiously and use just enough traction to get the job done.

Studs, also called caulks, are most commonly associated with eventing horses that often run and jump on wet grass or in the mud. Show jumpers sometimes also use studs. Removable studs screw into holes in the shoe before a run and

Studs

- Removable studs or caulks can be used as needed, with riders choosing the size and shape best suited to current riding conditions.
- It's natural for jumping horses to slide a bit on landing, so never use more traction than is necessary.
- The farrier needs to drill holes in the shoes to accommodate the studs.
- In between use, the holes must be plugged with rubber or cotton (soaked in WD-40 to prevent rust) to keep them from becoming clogged with dirt.

Borium

- A farrier can spot weld borium (tungsten carbide) on a horse's shoes to offer improved traction.
- Your farrier may recommend borium if you'll be encountering ice during winter riding.
- The borium can be applied in different ways, such as smears versus pearl drops, depending on the traction needed.

are then removed after the run. They work kind of like human cleats and come in different shapes and sizes depending on the terrain. Permanent studs can also be added by the farrier. The advantage of permanent studs is that the horse becomes accustomed to them, rather than suddenly having traction he's not used to. The disadvantage is that permanent studs cannot be adjusted based on the terrain, and the horse will have the added traction even when he doesn't need it.

MAKE IT EASY

If you ride in the snow, your horse's hooves may become packed with ice and snow, especially if he's shod, increasing his risk of slipping. For this reason, people often apply cooking spray to the bottom of the hooves to keep the snow from sticking, but this takes constant reapplication. Your farrier can apply snow pads that keep the snow from balling up.

Toe Grabs

- Toe grabs are ridges at the front of horseshoes.
- Toe grabs provide added traction and are most commonly associated with racehorses, where they're often used on the front feet.
- Studies of racehorses show that long toe grabs can cause injury, so many states have banned toe grabs longer than 4 millimeters for racehorses.
- Some barrel racers also use toe grabs for added traction.

Rim Shoes

- Rim shoes feature a rim or "swedge" around the bottom of the shoe that gives the horse added grip or traction.
- Even when the rim fills with dirt, the packed dirt offers better traction than a flat steel shoe.
- Polo ponies use rim shoes with a higher inside rim, and barrel racers commonly use rim shoes with a higher outside rim.
- Some racehorses are also shod with rim shoes.

CLEANING YOUR HORSE'S HOOVES

Use these techniques to make this essential care and maintenance task easier

Picking your horse's hooves may seem self-explanatory, but gaining your horse's full cooperation makes the job easier, especially considering you'll need to check his feet daily if possible and always before and after any exercise. Cleaning your horse's hooves removes rocks or other hard objects that could cause damage, removes manure and mud to help prevent thrush, and allows you to assess hoof health and catch problems early.

Most adult horses are used to having their hooves cleaned and handled, but that doesn't mean they're always cooperative. To gain your horse's cooperation, follow this method of hoof cleaning: First, establish an order for how you clean your

Position Yourself Safely

- To clean your horse's hooves, stand facing the opposite direction of him.
- If your horse is resistant to picking up his foot, you may need to lean your shoulder into him as you bend his leg and pull the hoof up.
- Point your toes away from the horse so that if he puts his foot down you will not get stepped on.
- Always be prepared to step aside if your horse moves suddenly toward you.

Cleaning the Hoof

- First, clean out the grooves on either side of the frog. If the dirt is hard packed, you may have to use a bit of pressure.
- Always push the hoof pick and dirt away from you.
- Gently clean the small groove in the center of the frog.
- Lastly, use the brush on the end of your hoof pick to brush away any remaining dirt and inspect the entire underside of the hoof for any problems.

horse's feet, such as left front, left hind, right front, right hind. This way the horse learns what hoof you'll ask for next and can prepare himself. Further prepare your horse for which hoof you want picked up by running your inside hand slowly down the back of his leg, from top to bottom. As you approach the fetlock area, bring the hand around the inside of his leg, palm flat against the front of the foot to scoop it up as you say the command, "Up." If he does not pick his hoof up, lean your shoulder into him to move his weight to his other side.

Once you have your horse's hoof up, he may try to snatch it away from you. If so, follow the motion of his leg forward and back, keeping the hoof up. Soon he'll realize he cannot take his hoof away and will hold it still. When you're done cleaning the hoof, set it down gently rather than dropping it. By following the above methods, your horse will soon be picking up his feet for you before you even ask!

Supporting the Hind Leg

- Because a horse's hind legs are heavy, many people choose to use the inside of their lower thigh, just above the knee to help support the hind leg.
- With your inside leg helping to support the horse's leg, use your left hand to cup and support his hoof.
- If you use your leg in this way, be prepared to quickly remove it and move out of the way should your horse kick or begin to struggle strongly.

Working with Older or Arthritic Horses

- Some older horses with arthritis issues or other leg problems have a hard time holding their hooves up high for cleaning and may find this position painful.
- If your horse has trouble holding his hooves up high, try to hold them lower to the ground to make hoof cleaning easier for him.
- Because this method can be hard on your back, try to perfect your technique and clean your horse's hooves quickly so that the process is better for both of you.

THRUSH

Thrush is a common problem for horse owners that can be managed with regular maintenance

Thrush is the most common hoof problem horse owners encounter. Some will only have to deal with it occasionally, while for others it's practically a constant battle. Thrush is a bacterial and fungal infection of the frog. You can spot it by its foul smell and black tarlike discharge, which often ends up on the end of your hoof pick as you clean out the collateral grooves along the frog or the small frog sulcus (groove) in the center of the frog. Thrush can affect all four feet.

Thrush is usually associated with wet conditions, and indeed a soggy pen that's not regularly cleaned makes for prime conditions. Regular maintenance, including keeping your horse's pen clean and picking his hooves out daily, is im-

Regular Hoof Cleaning

- Cleaning your horse's hooves daily allows you to check for the smelly black goo that's a sure sign of thrush, assess overall hoof health, and catch any concerns early.
- Daily cleaning also removes manure and other fungus- and bacteria-harboring debris that can lead to thrush.
- Be sure to thoroughly clean the collateral grooves along either side of the frog and to gently clean the frog sulcus, the small groove at the center of the frog.

Liquid Treatments

- There are many over-the-counter topical liquid thrush treatments available.
- Ask your farrier or vet if they have a brand of treatment they recommend.
- Read the directions carefully. Most topical remedies need to be applied daily to a clean hoof until the problem is remedied.
- Bleach has long been used as a home remedy for thrush but is not recommended and is too harsh to use on hooves.

perative. Your horse also needs at least one dry place to stand, even if he lives in a paddock or pasture. However, sometimes thrush can develop in even the best of conditions.

If you catch thrush early, it can usually be combated with the proper maintenance procedures combined with applications of over-the-counter thrush treatments, which are usually available at your local tack and feed store. Be sure your horse is on a regular trimming or shoeing program, at minimum every eight weeks. If the frog area is overgrown, it can create flaps and pockets that make treating thrush difficult.

If thrush is not treated early or the infection progresses too far, it can spread to the sensitive parts of the hoof. A horse that's sensitive to hoof cleaning is definitely telling you something. If thrush is not responding to treatment, or you suspect the infection has progressed too far, call your farrier out to evaluate the situation. You may also need to consult with your veterinarian to resolve the infection.

Packing Treatments

- Rather than a liquid treatment, your vet or farrier may recommend a hoof-packing treatment to combat thrush.
- Hoof-packing treatments are the consistency of thick tar or putty and can be packed around the frog.
- Hoof packing remedies have the benefit of helping to seal out debris, which can be especially helpful if you are not able to treat the hoof daily.

Living Conditions

- Thrush is a bacterial and fungal infection, so water alone won't lead to thrush.
- Wet conditions combined with a dirty pen make prime conditions for thrush.
- If your horse lives in a stall, keep the bedding clean and dry to help prevent thrush.
- If your horse lives outdoors, remove manure regularly and be sure he has dry places to stand, such as a three-sided shelter with rubber mats on the floor.

SAFE LIVING CONDITIONS

Proper pen maintenance is a key part of keeping your horse sound and healthy

Your horse spends most of his time in his pen, yet many owners don't give it much thought when considering the horse's soundness. However, his living conditions have one of the biggest impacts on his health and soundness.

Most experts agree that a more natural lifestyle, such as living outdoors in a paddock or pasture, is better for horses' mental and physical health. A horse that can move around in a large enclosure has better circulation in his legs and hooves, and is able to move his ligaments, tendons, and joints, which helps keep them healthy, rather than standing around getting stiff.

Stall-kept horses are also more likely to develop vices, like

Keep It Clean

- In the wild, horses do not need to stand or even step in their own manure.
- Forcing a horse to stand in manure and urine is not healthy for his hooves and can lead to problems.
- Stalled horses need soiled bedding removed daily and fresh bedding added. Twice daily cleaning or more is preferable, and daily cleaning is the minimum necessary. Stalls should be stripped (bedding removed and replaced) regularly.
- Horses kept outdoors also need manure removed regularly.

Fencing

- Unsafe or poorly maintained fencing can easily injure a horse's legs, so check fencing regularly.
- It's not uncommon for tendons, ligaments, or other soft tissues to become damaged by wire fencing or by protruding boards.
- Some fencing materials were meant for cattle and not for horses. Barbed wire, for example, is a very dangerous fencing for horses. Large-square wire fencing is also unsafe, as horses can place their hoof or leg through the squares.

stall weaving or pacing, which can be hard on hooves, joints, tendons, and ligaments. However, some horses must be kept in stalls, and for these horses, regular exercise and a clean and safe environment will help keep them happy and healthy.

Stall flooring needs to be relatively level and drain well. Some choose to use rubber mats on top of the floor. The horse also needs adequate padding to stand and lie down on. There are mats designed to take the place of bedding, and there are various bedding materials, including straw and shavings. One caveat with wood products such as shavings: Be sure the shavings are intended for horse use, as certain woods are harmful to horses, as are dusty shavings.

If your horse is kept outside, it's important to consider drainage. Even in rainy conditions, your horse needs dry places to stand. There are a variety of outdoor footing options for pens, such as decomposed granite and mulches (made out of horse-safe materials). Choose the best option depending on where you live and how large your horse's pen is. A soft place to lie down is also appreciated.

Hazardous Objects

- Remove sharp objects and rocks from the horse's pen and smooth areas that could cause tripping.
- If you have access to a metal detector or magnet, use it for locating stray nails or other sharp pieces of metal that may be lost in your horse's pasture or paddock.
- Equipment such as tractors should not be stored where horses have access to them, as a horse can easily injure himself on the sharp metal parts.

Pasture Buddies

- Horses are herd animals—social creatures that enjoy company.
- Your horse may be happiest with a pasture buddy, but keep in mind this increases his risk of injury, especially from kicks.
- If your horses are shod, remove their hind shoes before pasturing them together. This decreases the risk of severe damage from one horse kicking the other.
- Also reduce injury by introducing horses to each other slowly, giving them plenty of space and feeding them separately.

NUTRITION AND SUPPLEMENTS

Healthy legs and hooves require quality feed and a balanced diet

"You are what you eat" is largely true for humans and horses. If a horse is fed too little or poor quality protein, he won't produce adequate body proteins for healthy hooves, bones, tendons, and muscles. Hooves are constantly growing, and nutritional deficiencies show up in them. Horses with certain health issues, pregnant, growing, or elderly horses have special dietary needs. But a healthy adult horse's nutritional needs can often be met by good quality hay and a free choice salt lick.

The types of hays available and their quality vary depending on where you live. Hay falls into two categories: grass hays and legume hays. Grass hays include timothy, orchard, coastal Bermuda, bromegrass, fescue, and others. Legumes include alfalfa and clover. It's best to feed a grass and legume hay combination because grass hays are higher in fiber and lower in protein, while alfalfa, the most common legume hay, is slightly lower in fiber but can have too much protein. Grass

Pasture Grass

- Pasture grass can take the place of hay, but to meet 100 percent of your horse's needs, the pasture must be large enough (two to three acres per horse) and carefully maintained.
- When pasture grass is not plentiful enough to provide for your horse, supplement his diet with high-quality hay.
- Always introduce grazing slowly—a bit at a time—to avoid problems such as laminitis (see page 140).

Grass Hays

- In the wild, horses spend most of their day eating, so high-fiber grass hays help keep a horse happily munching without gaining too much weight.
- Talk to your vet about which grass hay is best for your horse depending on what's available in your area (timothy is pictured).
- Some horse owners offer grass hay "free choice," where the horse can munch on it whenever he wishes.
- Avoid hay high in weeds, excessively dusty hay, and moldy hay.

hays also tend to be lower in many nutrients, such as zinc, selenium, and vitamin E.

Grass and legume hay's nutrient values vary depending on the type of grass or legume hay, where it was grown, and at what stage and time of year the hay was cut. Therefore, consult your vet to create a feeding plan that works for your location and horse, and to see whether any additional feeds or concentrates are needed.

GREEN LIGHT

Biotin (a B vitamin) and zinc methionine are two of the most common ingredients in hoof supplements. Hoof supplements are intended to create stronger, healthier hooves. However, to avoid wasting money or providing your horse an overabundance of certain nutrients, consult your vet before adding any supplement to your horse's diet.

Alfalfa Hay

- Alfalfa is the most common legume hay.
- Quality hay should be bright green in color.
- The stalks of hay are mostly fiber, and the leaves contain the protein. Alfalfa is often very leafy and is higher in protein than grass hays.
- In addition to being higher in protein, alfalfa has a high calcium to phosphorous ratio, which is why it is often paired with a lower calcium grass hay.

Grains and Pelleted Feeds

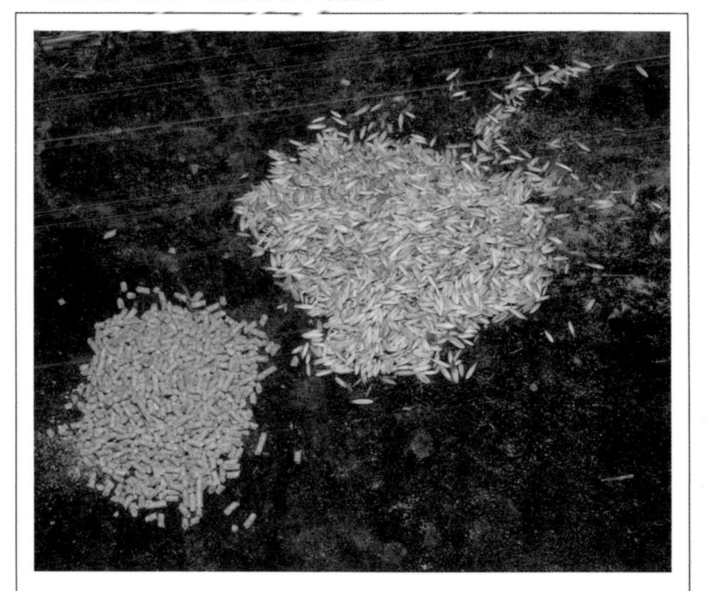

- Grain rations provide the horse with extra calories and energy.
- Unless a horse is in heavy training, he doesn't need an added grain ration.
- Besides grains, there are many pelleted and concentrated feeds on the market.
- There are even some bagged feeds, called complete feeds, created to be all the nutrition your horse needs. However, pelleted complete feeds are usually consumed quickly by the horse and won't meet his need for long-stem forage.

EXERCISE

Exercise can help keep a horse healthy or cause him injury, so it's important to condition him right

If you currently work out or used to participate in sports, then you already know a lot about conditioning. Many human conditioning principles are similar to those in the equine world. The important thing is to be mindful of your horse's condition to avoid injury.

Just as people warm up and cool down before and after every workout, so should your horse. A warmup usually consists of walking and then trotting, slowly increasing the intensity of the work to let your horse warm up—stretch and warm his muscles, increase his heart rate, etc. A cooldown is the opposite—work will slowly decrease in intensity, usually ending with a long, slow session of walking that allows the

Turnout

- If your horse is stalled, he is dependent on you to get him out of his stall every day for exercise or turnout.
- If you're not able to exercise your horse daily, he can move about and entertain himself if he has constant access to a pasture or paddock.
- Horses that live outdoors with room to roam will not condition themselves for work but are able to move about, which helps circulation in the lower legs, promotes healthier hooves, and prevents stiffness.

Slow Work

- Slow work over easy terrain is least concussive and stressful to a horse's joints, tendons, and ligaments.
- If you only ride once or twice a week, and your horse is not exercised between rides, he will only be conditioned to handle a short, slow ride.
- "Weekend warriors" that ride hard on the weekend and don't exercise their horses during the week are putting their horses at risk for injury.

heat buildup to dissipate. The horse's heart rate lowers, his respiration slows, and his sweat begins to dry.

At a minimum, exercise your horse three times a week. Horses in a serious conditioning program or working toward a performance goal are often worked five or six days a week.

A workout's intensity and length might vary from session to session. For example, you may take a day off from ring work to take a trail ride. Just don't ask for a great deal more intensity or length than your horse is conditioned to perform. For example, if your horse is accustomed to taking hour-long rides at the walk as his workout, adding five minutes of trotting or adding ten minutes in length shouldn't be a problem, but asking your horse to long trot for half an hour or adding an hour's length to your ride would be too much. Intensity and length have to be added gradually to build up a horse's condition.

How you ride affects your horse's balance, movement, and ultimately his soundness, so it's important that you ride in proper alignment. If you are a beginner, find a qualified instructor. Your horse will thank you!

Conditioning the Horse

- When conditioning your horse, it's important to build up his strength and endurance gradually to avoid injury.
- Never add difficulty and speed to your workout at the same time.
- For example, start your hill work at the walk, and only once your horse is conditioned for that, add a bit of trotting hill work.

Strenuous Work

- Athletes like jumpers, reiners, barrel racers, cutters, and dressage horses must be kept in excellent condition to perform their jobs.
- Conditioning doesn't mean you have to perform the most difficult tasks every day, however.
- For example, many jumpers are ridden on the flat on days between jumping workouts, and many barrel racers are conditioned with work at the trot. This can help extend an athlete's career.

ARENA FOOTING

Maintaining good arena footing can help your horse stay sound and injury free

If you ride in an arena, the footing plays an important role in your horse's soundness. Arena footing must provide cushion and appropriate traction. The ideal depth of footing varies slightly by discipline—between 2 and 4 inches.

If you've ever walked in deep sand, you know the problems created by too-deep footing. Footing 6 inches or deeper makes it hard to pull the foot back out and can cause injury to soft tissues, such as tendons and ligaments. However, footing that's too shallow or hard packed increases concussion on your horse's hooves and legs. Appropriate traction is not too slippery but allows for a natural amount of give when a horse lands from a jump or a reiner goes in for a sliding stop.

Proper Depth

- Ideal footing material depth is about 2 to 4 inches.
- It's common to have to add to your arena footing every couple of years.
- The footing in your arena will likely have to be completely redone and replaced every ten years or less, depending on the material used, how much traffic the area sees, and how well the arena is maintained.

Too-Deep Footing

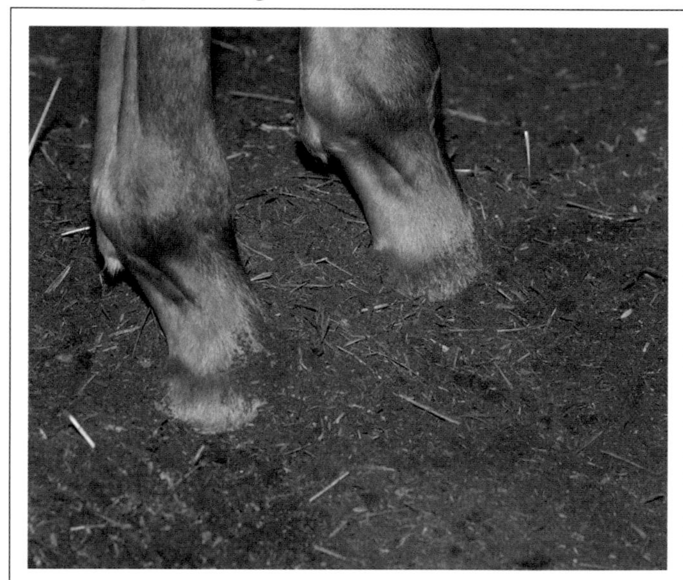

- Although ideal depth does vary by discipline, footing that's more than 5 inches deep will place added strain on your horse's legs.
- Think of how tired you get walking in deep sand on the beach, and you'll know how a horse feels when he's asked to work in arena footing that's too deep.
- Horses worked in too-deep footing will quickly tire, and continued work can lead to injury.

Arena footing must also be level and at an even depth in all areas, which requires constant maintenance and upkeep. Before footing is added, the area must be properly leveled and an adequate base layer created. After your arena footing is in place, additional footing material must be added as needed. Arenas are often watered to keep dust down, and must be regularly dragged, which involves pulling a spiked drag behind a small tractor. This helps smooth and level the footing.

The most common arena footing material is sand that has been cleaned of silt and clay to reduce dust and screened for large particles. There are many other arena footing options, including special additives that hold water and reduce dust, wood-based materials, and synthetic footing made of rubber components. These materials should be maintained according to manufacturer instructions. Whatever type of arena footing you choose, be sure it provides proper cushion and traction for your horse.

Too-Hard Footing

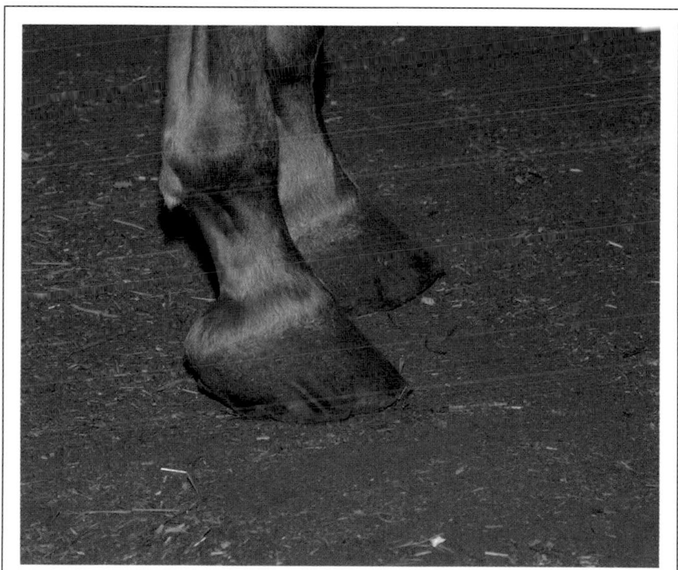

- An arena with too little cushion means your horse's legs and hooves are pounding on hard ground, adding concussion to his hooves and joints.
- Arena footing that is not properly maintained can become compacted in places, such as along the arena rail track, which is why arenas must be regularly leveled and dragged.
- Certain footing materials such as stone dust also have a great tendency to compact.

Arena Maintenance

- Dragging your arena is the most important aspect of arena maintenance, as it levels the footing and keeps it from compacting.
- How often you need to drag your arena depends on how much the arena is used.
- Arenas at commercial stables may need to be dragged more than once a day, whereas home arenas may only need to be dragged once a week.

USING BOOTS OR WRAPS

Boots and wraps can help protect your horse's legs but must be fitted and cared for properly

The main job for polo wraps and boots is to protect your horse's legs from interference injuries. Interference injuries happen when one hoof strikes another leg. This can cause splints or soft tissue injuries, such as damage to tendons or ligaments. Some horses interfere because of poor conformation or shoeing. Other horses that don't normally interfere may accidentally do so when performing demanding movements under saddle. Young horses just learning their balance and coordination under saddle are more prone to interfere and may knock a hoof into another leg.

Polo wraps must be smoothly and carefully applied. Boots generally come in small, medium, or large for front legs and

Boots on Trail

- Many trail riders use boots to protect their horses' legs since they travel over uneven terrain.
- If you're out on trail, remove your horse boots when riding through burrs, foxtails, or deep sand, which can get in the boots and cause the horse irritation.
- Remove boots when crossing water so that your horse doesn't have to ride for several hours in wet, soggy boots.

Types of Protection

- Horses need different types of protection depending on the work they do.
- Jumpers may use open-front boots, polo wraps, or other types of protection, depending on the horse's needs and the rider's preference.
- Some protective boots feature felt or fleece lining.
- Protective boots first became popular in the English riding disciplines, but now there are many Western specialty and general riding boots as well.

for hind legs. However, just as with human clothing, sizes vary, so it's best to call or e-mail the manufacturer or talk to a knowledgeable sales representative at your local tack store if you're not sure what size your horse needs. Boots come in many different styles, so read the manufacturer instructions on how to apply them correctly. Improperly applied boots can cause damage to the soft tissues of the leg. Boots that don't fit correctly or are misapplied can also cause chafing. In general, boots and wraps should be applied snugly and evenly, but not too tightly.

It's important to keep your boots and wraps clean and in good working order. After every use, check your horse boots for damage or cracks, and use a stiff brush to remove dirt and hair. Dirt and debris can cause leg irritation and chaffing. Clean and condition the outside of leather boots to keep them supple. Synthetic boots can usually be hosed off and dried in the shade to avoid sun damage.

Competition Boots

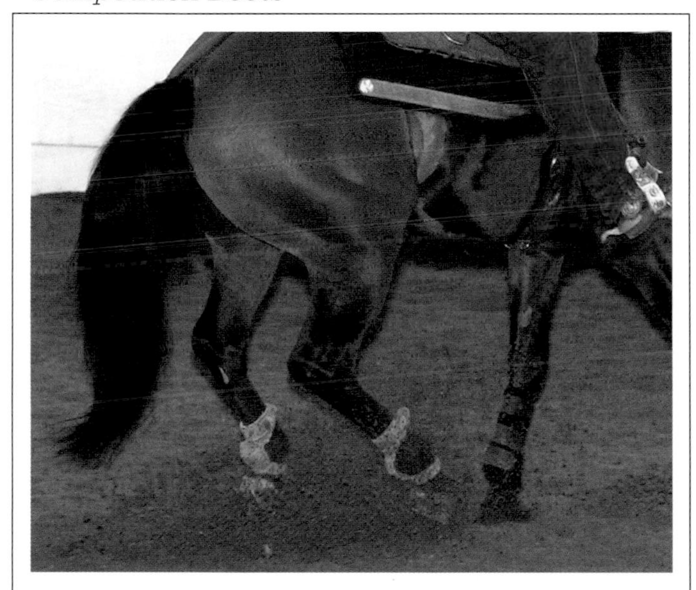

- Although leather is no longer the most common material for most leg boots, it is still the primary material for skid boots, which are used in reining and cutting.
- However, keep in mind that only certain disciplines allow horses to wear boots in competition. For example, most eventers wear some type of protective boots, but you won't see a Western pleasure horse wearing boots in the show ring. If you're not sure, check your competition rules.

Interference

- A horse's conformation, the way he's shod, and his maturity can affect his way of going and increase his likelihood of interfering, where one hoof strikes another leg.
- There are different types of boots for different parts of a horse's leg, and many people choose boots to protect the area where their horse is most likely to suffer an interference injury based on his way of going and the type of work he does.
- This horse is wearing bell boots to protect him from overreaching.

POLO WRAPS

Polo wraps take practice and careful attention to apply correctly

Polo wraps are made from cloth that's wrapped around the horse's leg for protection. They also work as leg warmers, helping to warm the horse's legs during workouts. Polo wraps were originally used on polo horses that not only had to worry about interference injuries but also hits from the mallet, ball, and other horses. Soon riders in other disciplines started using polo wraps, too.

Polo wraps or exercise bandages protect a large area—from below the horse's knee to his fetlock area. They're also relatively inexpensive compared to many boots. However, polo wraps take added time to roll and apply. Polo wraps also absorb urine, mud, and moisture, and shavings and other debris stick to them easily. For this reason, polo wraps are not a good choice for trail riding. Even for arena use, they require regular washing, usually after every workout. To wash your polo wraps, place them in a mesh laundry

Starting the Wrap

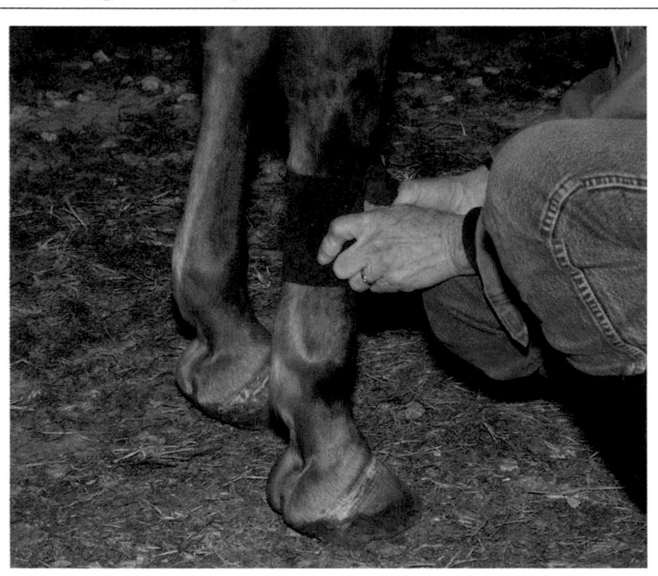

- Start the wrap on the cannon bone below the knee or toward the middle of the cannon bone (longer cannon bones or hind legs will require you start in the middle of the cannon bone to have enough wrap to finish).
- After the first pass, the wraps should be applied at a slant—about a 20 degree angle.

Wrapping Down

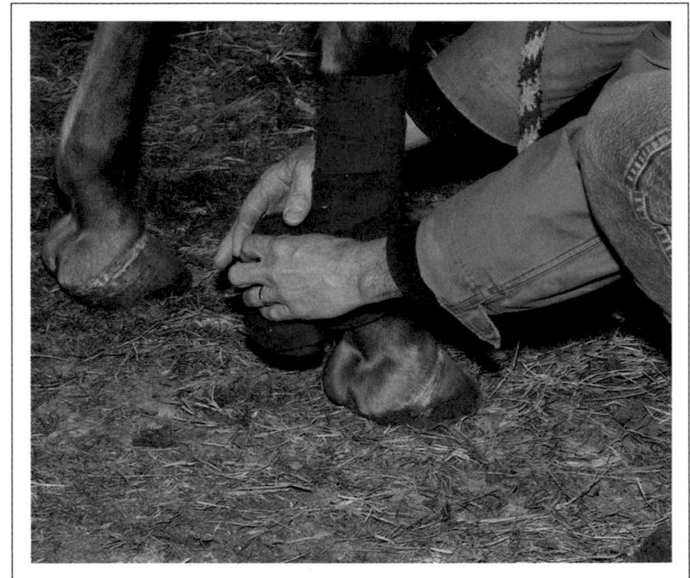

- Wrap down, overlapping each layer of the wrap halfway over the previous layer.
- Continue working your way down the horse's leg and around the fetlock area.
- During each pass of the wrap, apply gentle pressure when the wrap is at the outside of the leg by pulling it straight back toward your horse's tail before you bring it around the back of the leg.

bag so that they don't tangle and wind around the agitator of the washing machine.

When rolling a wrap, roll from the Velcro flap toward the Velcro attachment so that the Velcro ends up in the correct place when you apply them to the horse. Roll tight and smooth. When the Velcro and material starts to wear out, it's time to replace your wraps.

ZOOM

Generally, legs are wrapped clockwise on the right side and counterclockwise on the left. It's important to keep the wraps smooth. Any bumps or wrinkles in the material will place unwanted pressure on the legs. Apply the wraps snugly, but not too tight. You should be able to fit a finger underneath the finished product.

Wrapping Up

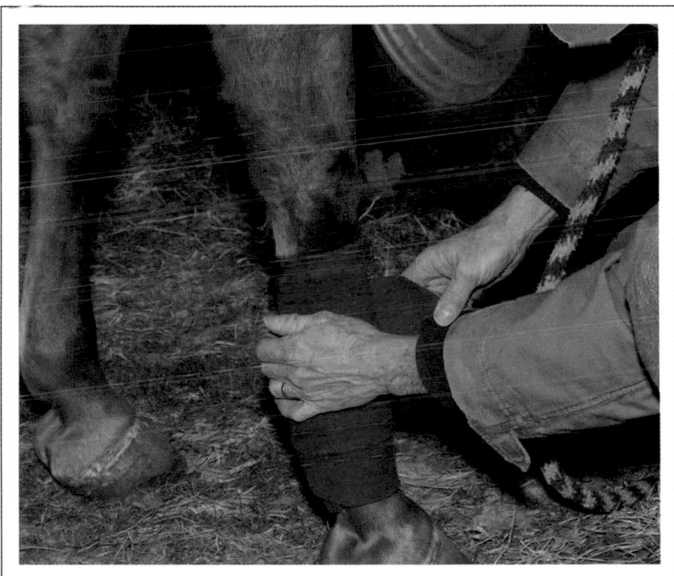

- After you apply the wrap to the fetlock, work your way back up the leg, again overlapping each layer of wrap halfway over the previous pass.
- If any wrinkles or bumps appear, go back and fix them before moving on.
- Applying polo wraps correctly takes practice.
- To avoid getting stepped on or kicked, don't position yourself behind your horse's leg and don't sit down.

Securing the Wrap

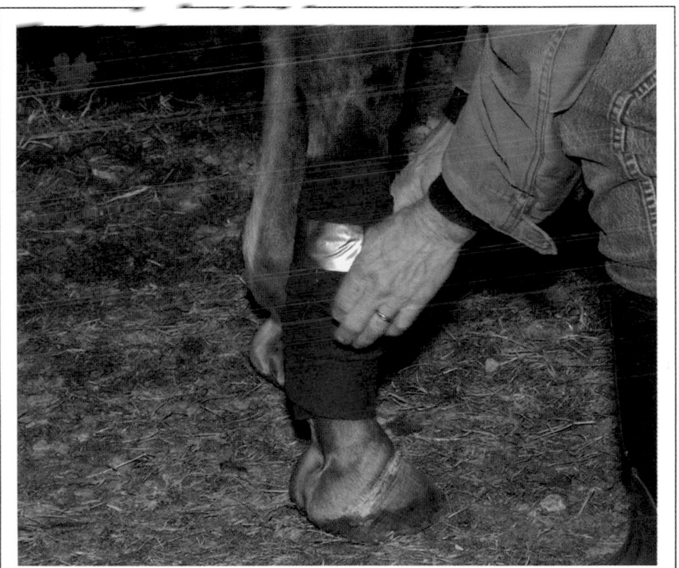

- The wrap should end on the cannon bone just below the horse's knee.
- Your last couple of passes of wrap should be applied straight rather than at an angle to make fastening easier.
- Smooth the Velcro flap over the Velcro attachment.
- To make sure your wrap stays in place, you can put a short piece of duct tape over the Velcro fastener.

RIDING BOOTS

Sport boots, splint boots, and bell boots are the most common general riding boots

Neoprene sport boots, splint boots, bell boots, and combination boots are general riding boots, meaning they were not created for any one particular sport but are used in a variety of riding disciplines. It's important to understand what these common boots can and cannot do for your horse.

Neoprene sport boots cover the leg from below the knee to around the fetlock. Because neoprene has some shock and concussion-absorbing abilities, sport boots can help protect the leg from interference injuries. Sport boots also have an added strap that wraps around the fetlock. The boots' design and material offer some support to the tendons and ligaments, but the amount of support a boot can provide is

Sport Boots

- As with most types of boots, sport boots come in small, medium, and large. Sport boots are also made for either front legs or hind legs.
- When putting on your horse's sport boots, slide them down into place to smooth the hair.
- The Velcro should be on the outside of the leg, and Velcro fastener should be applied snugly but not overly tight.
- Always read manufacturer instructions for fit and application.

Splint Boots

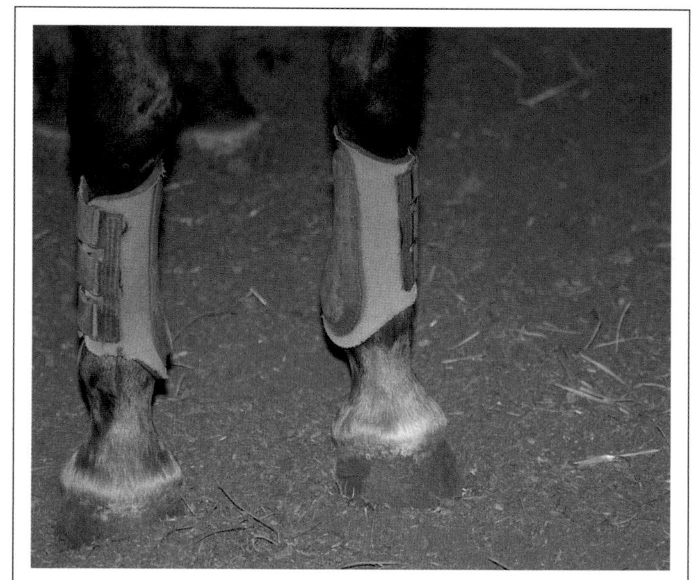

- Splint boots come in a variety of materials, including leather, neoprene, rubber, and combination hybrids.
- Splint boots are usually available in small, medium, and large.
- The thick padded side of the splint boots slides into place along the inside of the cannon bone, with the wider cupped part fitting over the upper fetlock area.
- Splint boots often have Velcro closures—these should be on the outside of the leg, with the Velcro tabs tightened by pulling snugly back toward the horse's tail.

limited when you consider the sheer weight and force that's applied to the suspensory apparatus during movement.

Splint boots are so called because they help prevent splints—calcified lumps formed after knocks to the splint bones. Splint boots protect the inner part of the leg from below the knee to the fetlock from interference injuries.

Bell boots protect the coronet and heel bulbs from forging or overreaching, where the horse's hind hoof strikes the back of the front foot. They are also useful for horses wearing shoes that extend beyond the heel, such as egg bar shoes.

RED LIGHT

If you've ever removed a boot from a horse's leg after a workout, you know that the leg is generally hot and sweaty underneath. Because boots can make the leg hot, never leave them on your horse between classes at a show or if he's standing around between workouts. After a workout, cold hose your horse's legs to reduce their temperature.

Bell Boots

- Bell boots come in a variety of styles and materials.
- Some bell boots have scalloped pieces of plastic, others are solid and must be pulled on. The most common kind are fastened with Velcro or a buckle around the lower pastern.
- Bell boots can be made of rubber or neoprene. Neoprene stays in place better.
- If you use bell boots, be sure they don't rub or irritate your horse's skin with continued use.

Combination Boots

- Combination boots combine a bell boot with a sports boot to offer a greater area of protection from interference injuries.
- Combination boots can be handy for young horses that need a great deal of coverage from interference injuries as they gain balance and coordination during training.
- Boots aren't just for riding: Horses can also benefit from leg protection such as combination boots when they're being lunged or turned out to play under supervision.

DISCIPLINE-SPECIFIC BOOTS

Specialty boots can offer specific protection for performance horses in Western and English disciplines

In addition to general riding boots, a number of boots, designed for specific purposes and disciplines, offer protection only where the horse needs it.

Both dressage boots and galloping boots are types of brushing boots that protect the horse's leg from interference injuries. They protect roughly the same area as splint boots—the inside of the leg from below the knee to the fetlock. Brushing boots generally have a tough outer material with felt or fleece lining the insides. Dressage horses may suffer interference injuries when performing the lateral work common to the sport. Galloping boots help protect the legs during fast work, such as the cross-country phase of eventing.

Dressage Boots

- If you've ever watched a dressage horse move laterally across the arena, crossing his front and hind legs, you'll understand that sometimes these moves cause interference injuries, where one hoof strikes the opposite leg.
- Dressage boots usually combine a tough outer shell for protection with comfortable inner fleece lining.

Open-Front Boots

- Open-front boots like these are often used by jumpers.
- Open-front boots help protect the horse from splints and interference injuries.
- The front of the boot is left open so that the horse can "feel" the jump rail and learn to jump carefully.
- With all the protective legwear options available, you can ask your horse trainer or riding instructor about what's best for your particular horse and needs.

Open-front boots were designed specifically for jumpers. They provide protection for the back part of the legs and partial protection for the sides of the front legs but are open in front so that if the horse rubs a rail when jumping, he can feel it. The theory is that by feeling the rails in this way, the horse learns to jump more carefully and cleanly.

Leg boots do increase the temperature of the leg, so some riders choose to only protect the fetlock/ankle area, which includes the proximal sesamoid bones. Fetlock/ankle boots are most commonly used for galloping and jumping, such as by eventing riders that may be using studs in the horse's shoes. Ankle/fetlock boots may also be needed if the horse travels too close behind, causing interference.

Skid boots protect the rear fetlocks of reining and cutting horses that may do sliding stops, spins, and quick turns that chafe or rub the hind fetlocks.

Note that before applying any boot, you should be sure your horse's legs are clean.

Ankle/Fetlock Boots

- Ankle/fetlock boots protect the inner fetlock area.
- A "cup" covers the inner fetlock, with a Velcro or buckle closure on the outside of the leg.
- Young horses may benefit from fetlock boots until their coordination and balance under saddle improve.
- Ankle boots are available in a variety of sizes and materials.

Skid Boots

- One of reining's signature moves is the sliding stop, where the horse's hind legs come underneath him, and he literally slides to a stop.
- Sliding stops, along with the spins and sharp turns common in reining and cutting, can chafe or rub the hind fetlocks. Skid boots protect this area.
- Some trainers only use skid boots if the horse's longer fetlock hairs get rubbed off.
- It's important to check fetlock boots often for dirt and debris buildup.

NONRIDING BOOTS

Boots can provide valuable protection for your horse even when you're not riding

There are a number of protective boots designed for use in the horse trailer to prevent injury or to use in the horse's corral to prevent sores.

Horses should not be hauled in a horse trailer without shipping boots, which offer leg protection. Because a horse can't see the upcoming turns or stops, and because horse trailer rides aren't smooth, horses often have to scramble to keep their balance. While scrambling, it's common for them to injure their legs. Some horses also kick the sides of the trailer or each other during transit.

Before shipping boots were popular, people frequently wrapped their horse's legs with standing wraps and applied

Shipping Boots

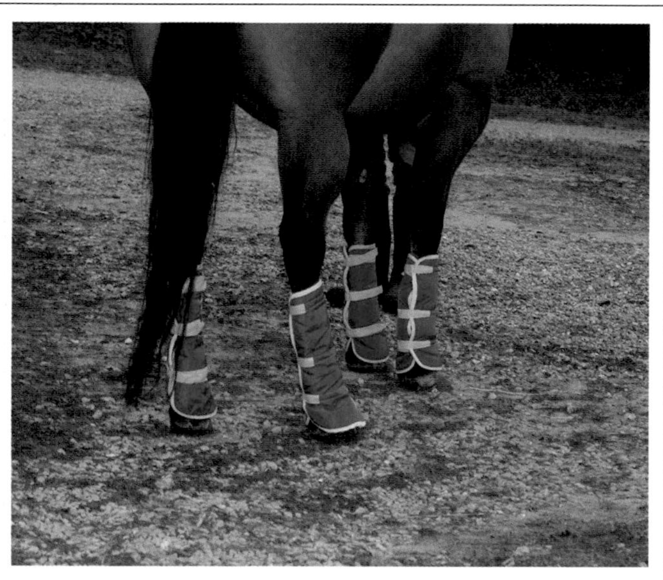

- Whenever a horse enters a trailer, he should be outfitted with protection on all four legs.
- To apply shipping boots, slide them into place to smooth the hair, and make sure the Velcro is on the outside.
- Shipping boots should be fitted snugly enough not to turn.
- Shavings and other debris easily stick to the fleece lining, so be sure boots are kept clean.

Hock Boots

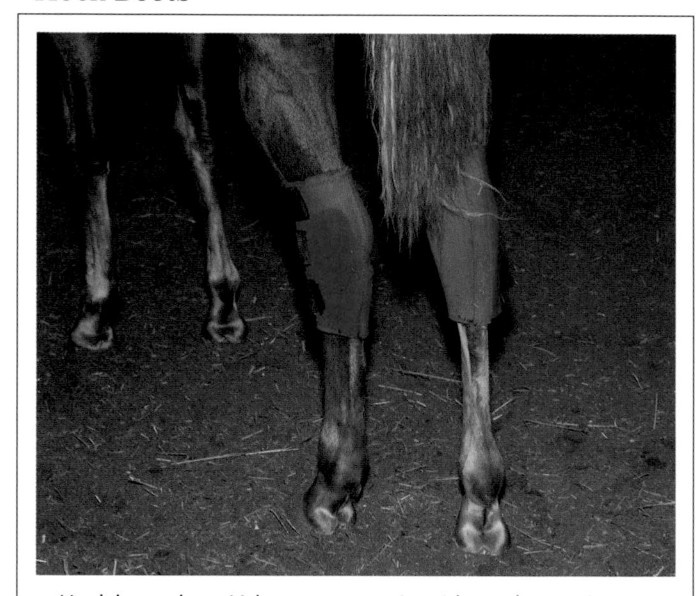

- Hock boots have Velcro around the hock and can be used for protection in the trailer or used in the stable to help prevent hock sores or to allow them to heal.
- Hock sores are common when the horse lies down on hard ground.
- As with any boot, it's important to check it regularly to make sure it hasn't slipped out of place and that no debris has gathered inside it.

bell boots, and even hock or knee boots. This is still an option, but secure the Velcro on the wraps with a piece of duct tape so that they don't unravel. Shipping boots are certainly handier. They're heavily padded with a canvas type outer layer and fleece on the inside, and they most often cover from the horse's knees and hocks down past his coronary band.

Knee boots are commonly used by Standardbred racehorses that suffer knee interference injuries. Knee boots can also be used in the horse trailer for knee protection.

Bed sore boots are designed to prevent bed sores on the fronts of the fetlocks and allow existing sores to heel. Bed sores can occur when horses lay down on hard ground, rubbing the fronts of their fetlocks raw. Bed sore boots cover the front of the fetlock and attach around the pastern below and the mid–cannon bone area above.

Shoe boil boots, or donuts, are used on horses with capped elbows—the swollen bump you often see on a horse's elbow.

Hock Sores

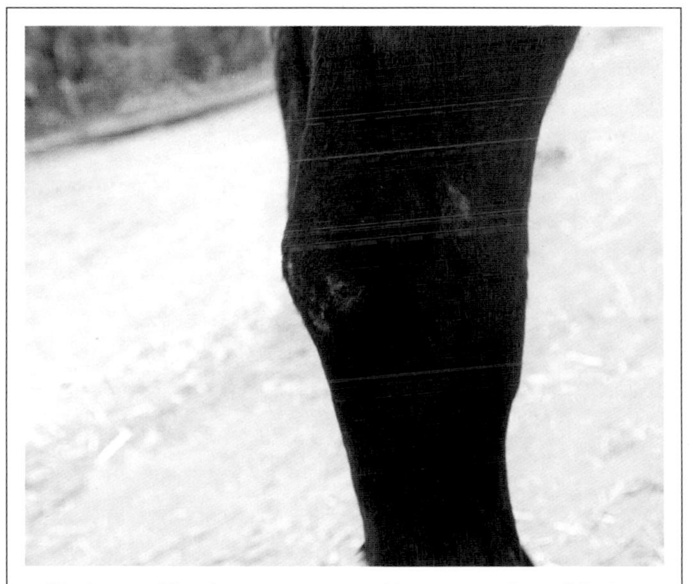

- Hock sores like these are common, especially for horses kept on hard ground.
- Generally, hock sores will not cause lameness unless they become infected or large and extremely sore.
- You can apply antibiotic ointment to existing sores to help prevent infection, but unfortunately ointments often attract dirt.
- Hock boots and softer ground/bedding can help prevent hock sores and allow existing sores to heal.

Shoe Boil Boots

- Capped elbows can be caused by the horse's shoe or hoof hitting his elbow when he lies down.
- A big rubber boot—called a "donut" or shoe boil boot—fits around the horse's pastern.
- When the horse lies down, the boot keeps his hoof away from his elbow.
- Check shoe boil boots and other boots left on your horse daily to make sure they're not causing chafing.

IS MY HORSE OFF?

Even subtle changes in movement and behavior can signal lameness

A "lame" horse is one that's not bearing weight or moving normally on one or more legs, signaling a problem. It's important to catch lameness early so that you don't continue to work the horse and make the problem worse, and so that treatment can begin before the condition progresses.

Sometimes it's very obvious a horse is lame. In the most severe cases he is "three-legged lame" or not bearing weight on one leg. Or he may be noticeably limping. Usually, however, the signs are subtler—something may just seem off about the way the horse is traveling. In a front limb lameness, a slight head bob, where the horse brings his head up and down, may be apparent when the affected limb is used.

Observation While Lunging

- Lunging the horse is a good way to assess his way of going, and some lamenesses are more apparent when the horse travels in a circle or goes in one direction versus the other.
- Is the horse's head bobbing up and down more than usual in response to weight bearing on one leg? This is a common sign of lameness.
- Is one hoof breaking over, or being lifted off the ground and set down differently than the opposite hoof? This is another indication of a problem.

Observation under Saddle

- A lame horse often won't feel quite right when you ride him.
- A mildly lame horse may feel like he's walking stiffly or taking a shorter step on one leg.
- The walk and trot are even gaits—the walk is a four-beat gait and the trot a two-beat gait—so this is where you'll notice any unevenness in movement.
- If your horse seems off, don't continue to work him. Dismount and observe him from the ground.

There are other times when you can't really see lameness at all, but the horse is telling you something. A sore horse may seem crankier than usual. He may pin his ears or wring his tail when asked for certain gaits or maneuvers under saddle. He may refuse to take a lead at the lope or canter or not want to move out or extend his trot. If these behavior changes continue, they're worth investigating. *Investigating* is truly the correct word, too, as getting to the bottom of things can be difficult.

When mysterious behavior changes like those mentioned above occur, you should first rule out obvious causes, such as ill-fitting tack—especially a poorly fitting saddle. If pain isn't visible to the eye or you don't see a reaction when you touch the horse, it can be hard to discover where the pain comes from. The source of pain could be in the legs or hooves, or it could be in the hips, back, or shoulders. However, keep in mind that horses don't usually become cranky or resistant without a reason.

Trot Test

- To assess whether any lameness is present when your horse moves, it can be helpful to ask a friend to trot the horse in a straight line on flat ground.
- The trot is often the gait where lameness is easiest to see, as it's faster than the walk but still an even gait.
- If there's any head bobbing present, it will usually be most visible at the trot.

Behavior Changes

- General behavior changes during workouts, such as crankiness, bucking, rearing, or refusals can indicate the horse is in pain.
- One example of a behavior change that often indicates a leg or hoof problem is reluctance to take one lead at the canter or lope.
- Reluctance to take a lead can be a sign that there's a problem in the outside hind leg, which is the leg that must push off to initiate an inside lead.

PHYSICAL SIGNS OF LAMENESS

Hot, swollen legs, pain, wounds, and an elevated pulse are all signs of trouble

Aside from the horse moving differently or limping, physical signs of lameness may also be detected when examining the leg.

Wounds are generally easy to spot if they break the skin. However, wounds that don't break skin can be harder to see but often create swelling. Small, superficial cuts and scrapes generally won't cause lameness unless they become infected. Deeper wounds or those that cause bone injury, bruising, or soft tissue damage can certainly produce lameness.

If you suspect your horse may be lame, use your fingertips to feel and examine the leg. If a horse consistently reacts to touch he normally wouldn't (or touch he doesn't react to on

Detecting Heat or Pain

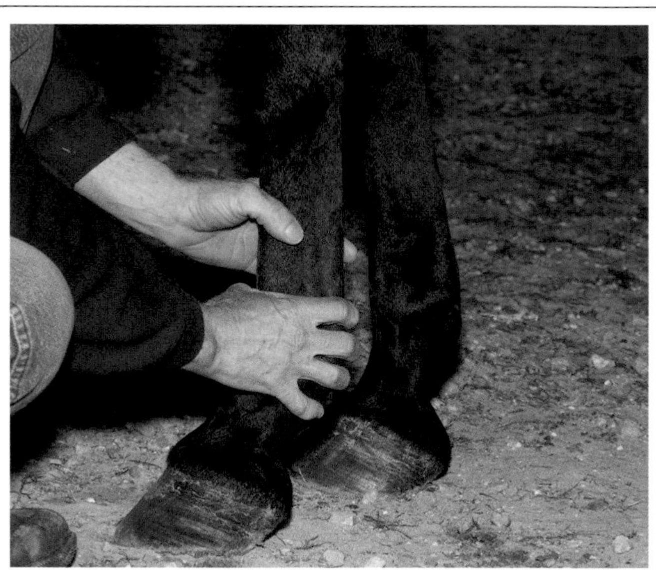

- When you clean your horse's pen, groom him, and clean his feet, visually inspect his legs and note his reactions to your touch.
- If your horse's legs look different than usual, or he reacts differently when you groom a leg or ask him to pick up his hoof, this warrants further investigation.
- Feeling the horse's legs with your fingertips and comparing both front or both hind legs helps you detect pain or heat, which are both indications of trouble.

Checking Digital Pulse

- To feel your horse's digital pulse, squat down beside his lower leg, and use your fingertips to feel around the back part of the fetlock, at the level of the proximal sesamoid bones or in the mid-pastern area, as shown in the photograph above.
- Feel for a cordlike bundle, which includes a vein, artery, and nerve.
- Once you locate the bundle, press your fingertips against it with varying pressure for several seconds, trying to feel the pulse.

the opposite leg), it's a good indication he's experiencing pain in that area.

Palpate or feel the suspect leg, looking for signs of heat or swelling. The best way to detect heat or swelling in a leg is to compare it to the same leg on the opposite side. Although heat and swelling can affect more than one leg, many lameness issues caused by injury or trauma only affect one leg. Place the fingertips of one hand on one leg and the fingertips of the other hand on the opposite leg in the same place. Does one leg feel warmer than the other? Next, visually assess the two legs, comparing the same parts of each leg to see if swelling is present.

In a healthy leg, the digital pulse in the fetlock area will barely be detectible. If the digital pulse is pounding and easily felt, then it's likely a sign of trouble. Check the digital pulse on all legs for comparison. Unlike pulses taken in other areas, you're not counting beats, you're checking for the strength of the pulse. It's helpful to check your horse's digital pulse regularly so that if there ever is trouble, you'll know the difference.

Swollen Legs

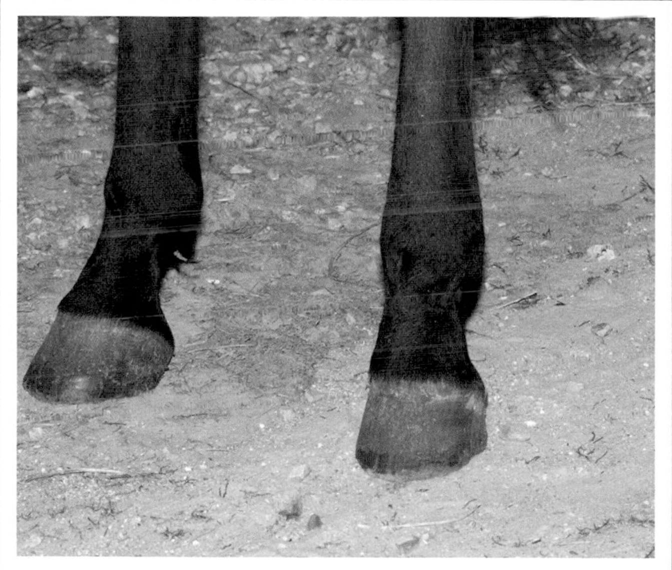

- It's important to be familiar with your horse's legs and what's normal for them.
- If the two front legs or two hind legs don't match, and one seems puffier in parts than the other, that's a good indication one leg is swollen, which indicates a problem.
- Sometimes both front legs, both hind legs, or all four legs become "stocked up" or swollen.
- Knowing your horse's legs well helps you catch changes early.

Wounds

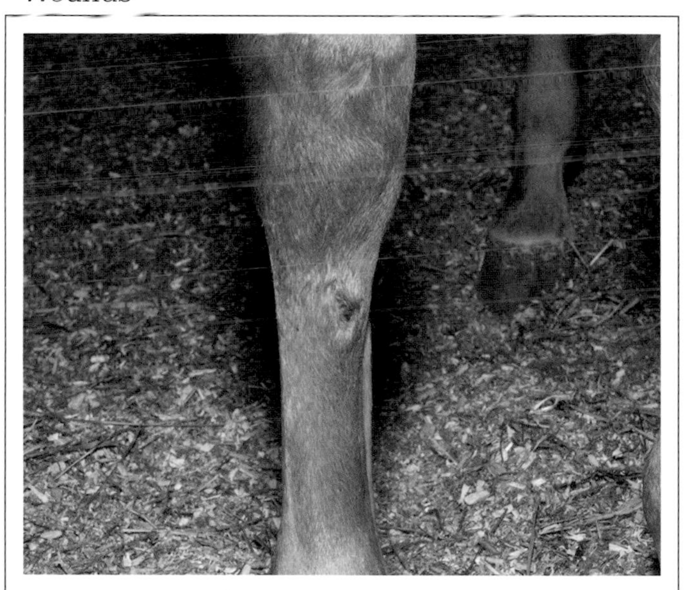

- Legs have very little padding, and tendons, ligaments, and bones are not far from the surface.
- Legs can easily be wounded by interference injuries, kicks from other horses, or run-ins with the stall wall, paddock fencing, feeders, or other objects.
- Because equine legs are so vulnerable, leg wounds can cause lameness, depending on their placement, depth, width, and whether they heal without complication.

WHEN TO CALL THE VET

Certain conditions always warrant a vet call, but when in doubt, play it safe

When it comes to calling the vet, it's better to be safe than sorry. If you aren't sure or you don't know how to handle a situation, call the vet. Before doing so, gather as much information as you can about your horse's condition so that you can clearly convey the problem. Once you have a working relationship with a vet, many times the vet can determine over the phone whether a visit is truly necessary or whether you can handle matters on your own.

Although a lame horse's vital signs may be normal, severe pain, and certain conditions like infection and tying up, can register as changes in a horse's vital signs. Anytime there's a problem or suspected problem, it's a good idea to take your

Wounds

- Sutures must be applied within six hours for best results, so call your vet if you think sutures may be needed.
- Even if a wound can't be sutured, the vet can thoroughly clean it and tell you how to care for the particular wound as it heals.
- Deep wounds or wounds that may have damaged joints, tendons, ligaments, or bone also require immediate veterinary attention.

Lameness

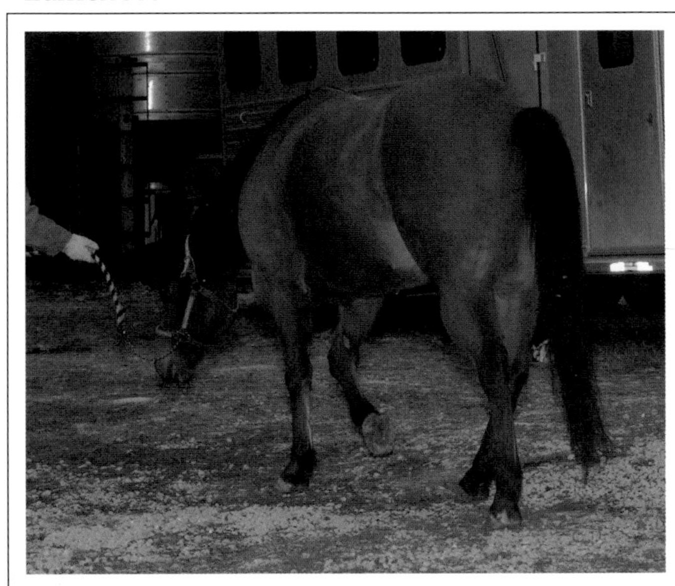

- Any sudden lameness, which may be accompanied by heat and swelling in the leg, requires veterinary attention.
- Call the vet immediately if your horse suffers a puncture wound to the hoof, such as stepping on a nail or other sharp object.
- You should also call the vet if you suspect a hoof abscess (see page 166) or laminitis (see page 140).

horse's vital signs. That way you can give your vet a complete picture if you need to call for a consultation. How to take a horse's vital signs, including his temperature, respiratory rate, heart rate, and how to check for dehydration are detailed on pages 212–213.

Even if vital signs are normal, any time a horse is obviously lame (not bearing weight or traveling normally) on one or more hooves, it's best to call the vet. Lameness is often accompanied by heat or swelling. Also consult your vet if your horse has a strong digital pulse or is reluctant to move or eat.

If you suspect your horse has fractured or dislocated part of his leg, this is also a veterinary emergency.

Puncture wounds or other serious wounds, including those near tendons, ligaments, or joints, need veterinary consultation to avoid infection and ensure proper healing. Wounds that won't stop bleeding also require immediate veterinary attention, as do wounds that need stitches. We'll discuss cleaning and care of wounds in a later section. Also see page 210 for what to have in your first-aid kit.

Behavior Changes

- Although behavior changes, such as crankiness or refusal under saddle, aren't veterinary emergencies, they do warrant veterinary attention if they continue.
- Behavior changes during training can be caused by ill-fitting tack, poor riding or training practices, and pain or lameness.
- A veterinarian can rule out pain or lameness as a cause for behavior problems, or if pain is the cause, the vet can help develop a treatment plan.

Not Responding to Treatment

- If you've been treating what you thought was a minor problem on your own, such as a stocked up (swollen) leg, and your horse is not responding to your treatment, consult the vet.
- Waiting too long to seek advice from your vet can make treatment more difficult.
- Mistreating problems on your own can exacerbate the condition.

SOUNDNESS EXAM

Your vet will utilize a variety of methods to determine where your horse's lameness comes from

If your horse has soundness issues, your vet will perform a thorough soundness exam. Soundness exams are also part of prepurchase exams, in which a horse is evaluated before purchase. Even if the location of lameness seems obvious, there may be less visible primary or secondary issues the vet must check for.

Hoof area problems, including abscesses, bruises, and navicular syndrome, are common, so your vet may begin his or her exam at the hooves. If a problem in the hoof area is suspected, the vet will then use hoof testers, which are big metal grips, to apply varying degrees of pressure to different parts of the hoof to check for underlying pain.

Palpation

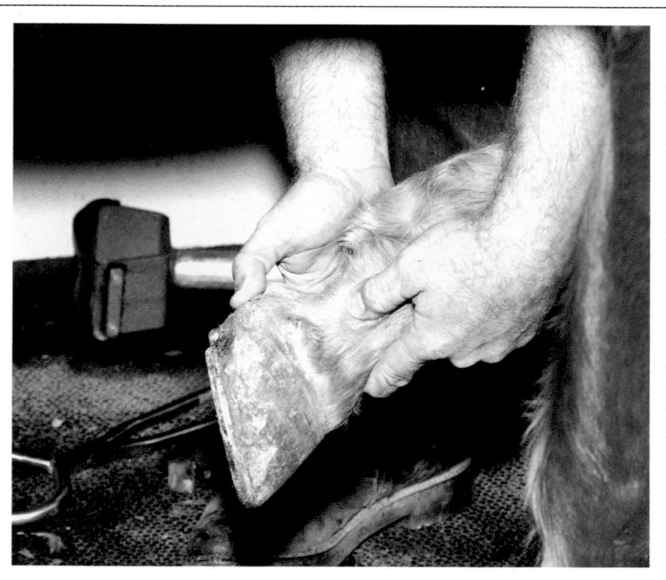

- A veterinarian is highly educated in horse anatomy and will carefully feel each of the horse's legs for heat, swelling, joint, bone, and soft tissue changes.
- The vet will likely palpate the leg as the horse stands and then again with the hoof held up.
- Palpating the leg in different positions allows the vet to feel and manipulate the various parts of the leg in different ways.

Flexion Tests

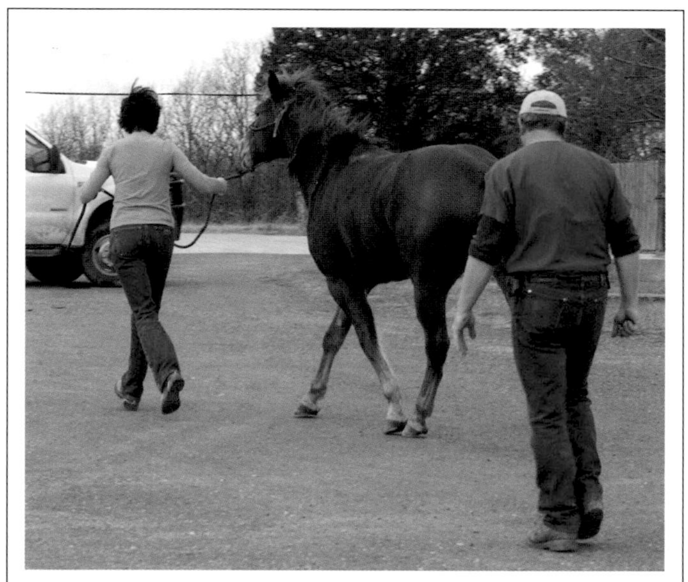

- After the vet flexes a limb, the vet will ask you to trot the horse in a straight line away from him or her to assess how the horse reacted.
- The vet will watch to see if the horse trots out lame and for how many steps the lameness lasts.
- If the horse resists flexion or tries to pull the limb away from the vet, it can also be an indication of pain in that joint.

The vet will thoroughly palpate the horse's legs. Palpating means the vet is feeling the various parts of the leg with his or her fingers. The vet will feel for heat as well as swelling and changes in texture so that he or she can detect tendon, ligament, joint, or bone changes. The vet will also manipulate the joints to assess their mobility and range of motion.

During a soundness exam, the vet will likely ask you to hand trot or lunge your horse so that he or she can assess the horse's movement, check for lameness, and evaluate where the problem may be stemming from.

The vet may also perform flexion tests, where the vet flexes a joint and holds it for thirty to ninety seconds and then asks you to trot the horse. How the horse trots after flexion can be an indication of pain or problems in the joint that was flexed. However, how long the joint was flexed for and how much pressure was applied affects the outcome, so other means of evaluation should also be used.

Nerve Blocks

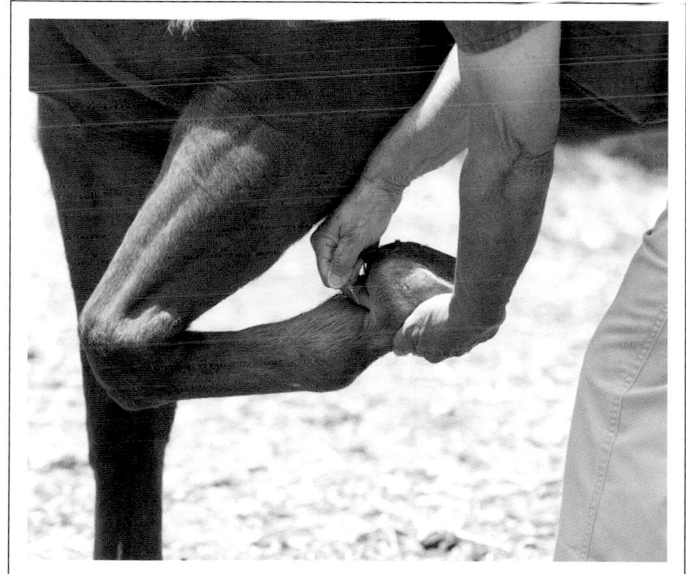

- Nerve blocks are a common and useful tool for the vet to determine where in the leg or hoof a horse is lame.
- With a nerve block, the vet blocks or numbs certain nerves, usually starting low on the horse's leg and working up until the affected area is blocked.
- If a horse's lameness improves after the nerve block, it's a good indication of what area the lameness is stemming from.

X-Rays

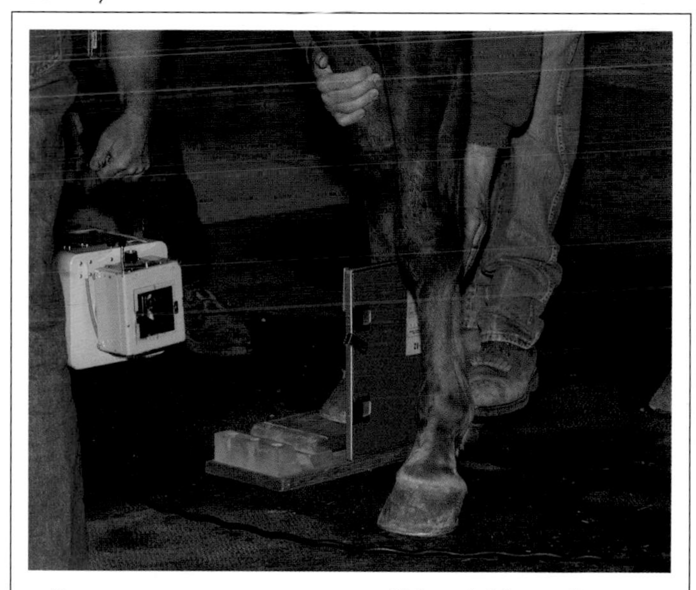

- X-rays are a common diagnostic tool used to determine the site and cause of lameness.
- Unlike many advanced diagnostics, x-ray equipment is portable and can be used on location during the lameness exam.
- Although it's usually difficult or impossible to determine soft tissue problems via x-ray, x-rays are excellent for showing bony changes or problems, such as bone spurs or ringbone.
- When x-raying the hoof, the vet may prefer to remove your horse's shoe.

ADVANCED DIAGNOSTICS

Advanced diagnostic tools help the vet determine the location and nature of a soundness issue

For soundness problems that can't be properly diagnosed with a routine soundness exam, or for ongoing soundness issues that are not responsive to treatment, your vet may recommend advanced diagnostics. Most advanced diagnostics require a trip to the vet hospital, as the equipment is not as portable as x-ray equipment. Your vet may also need to use more than one diagnostic tool. Advanced diagnostics can be costly but are often quite helpful in fully revealing the condition in order to create an effective treatment plan.

Although you've no doubt heard of arthroscopic surgery, arthroscopy was first used as a diagnostic tool and is still very valuable in this role. In arthroscopy, a tiny arthroscope (a flex-

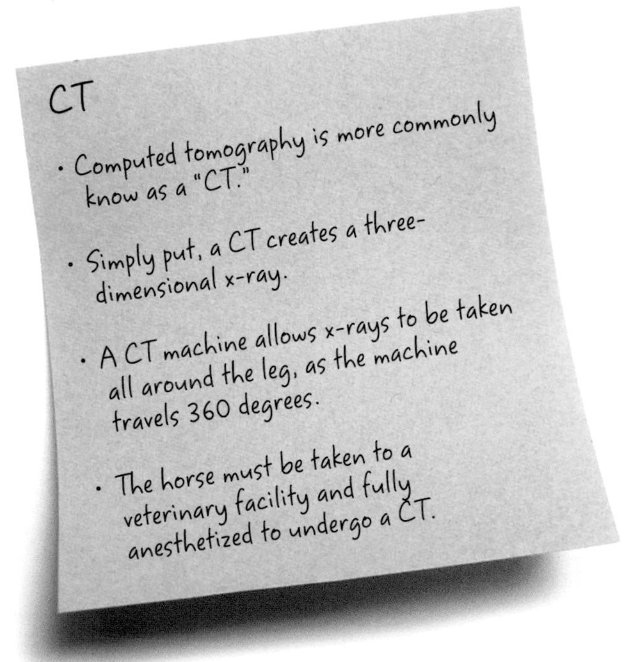

Thermography

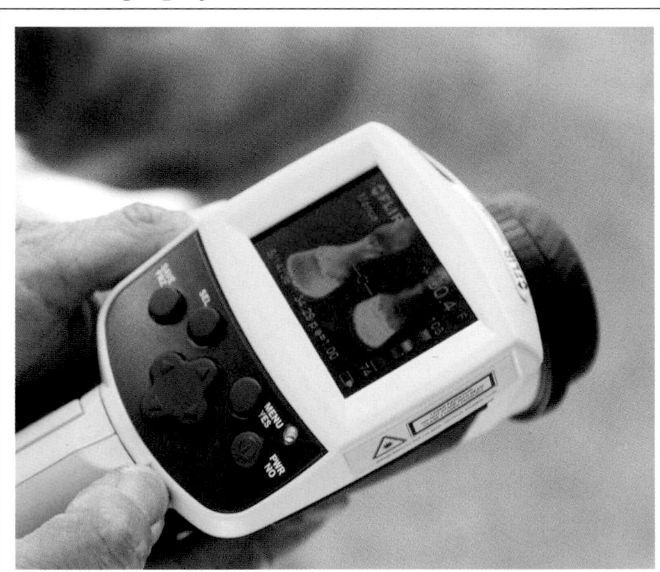

- Thermography is thermal imaging that picks up infrared radiation that is naturally emitted from the body to show variations in temperature.
- With thermography, the vet can compare the temperatures and circulation in the legs to see the extent of a problem or its location.

ible, fiber-optic scope) allows the vet to examine joints more carefully, including checking for changes in the synovial membrane and damage to the bone or articular cartilage.

MRI stands for magnetic resonance imaging and is an extremely valuable advanced diagnostic for a variety of purposes. MRIs produce detailed images of bone, soft tissue, and cartilage, so they are very helpful. In the past, MRIs were difficult to perform on horses, as the horse had to be laid down and put under full anesthesia. New standing MRIs only require light sedation and are becoming available in more areas.

ZOOM

In addition to advanced diagnostics, even the traditional x-ray is going high-tech. Many veterinarians are now using digital x-rays. With digital x-rays, you don't have to wait for the film to be developed; you can view clearer x-rays on high-resolution screens almost immediately.

Ultrasound

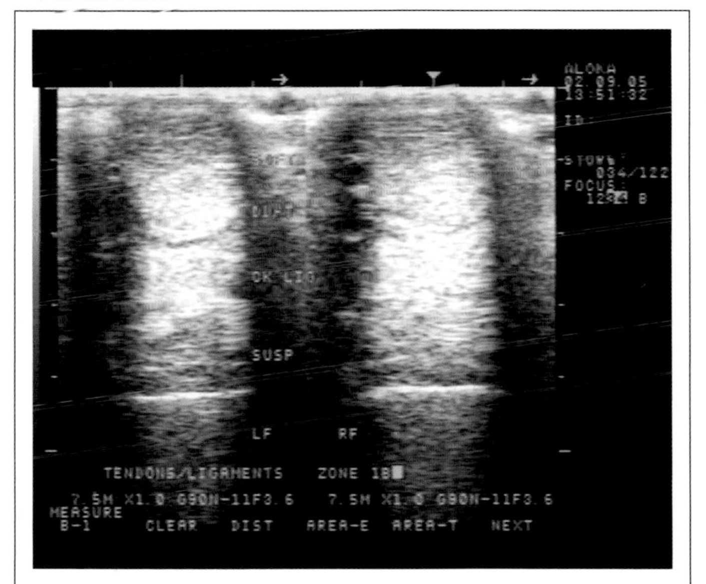

- Ultrasonography or sonography is commonly known as the ultrasound.
- Ultrasound uses sound waves to create visuals of soft tissues.
- The ultrasound technician applies a water-based gel on the area to be examined and then runs a hand-held transducer over the area. The resulting images will immediately appear on the screen.
- Ultrasound can show soft tissues problems, such as ligament or tendon injuries.

Nuclear Scintigraphy

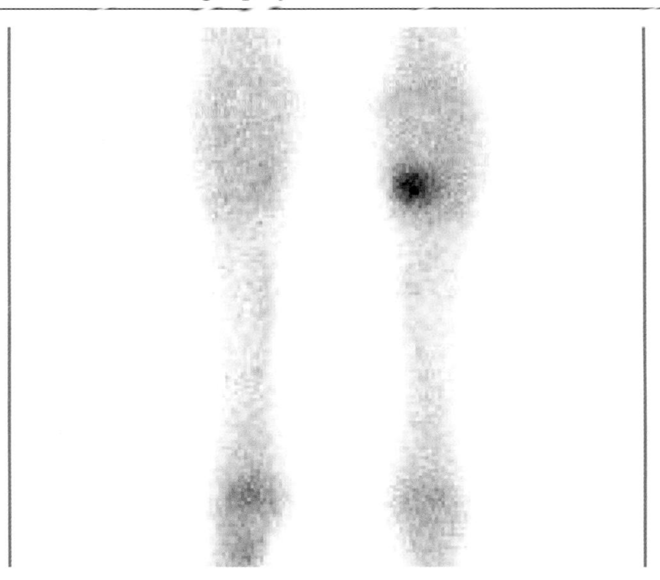

- In nuclear scintigraphy (commonly called a bone scan), benign radioactive substances are injected intravenously into the horse and then scanned.
- The radiopharmaceuticals are usually concentrated in "hot spots," where they are more readily absorbed, such as where there is increased blood flow due to inflammation or an increase in bone-forming cells.
- In this way, nuclear medicine is helpful in showing where a problem is located. It's important to image more than one leg for comparison.

LEVELS OF SOUNDNESS

Many soundness issues can be managed or treated to keep the horse comfortable and usable

When veterinarians assess a horse for soundness, they usually grade any lameness they find on a scale from one to five, with zero meaning no lameness is perceptible under any circumstances. A grade one lameness is not easily observable and not consistently apparent; a grade two may not be consistent at the walk or trot in a straight line but is consistent in other circumstances, such as circles or under saddle; a grade three lameness is consistently observed at the trot regardless of circumstances; a grade four lameness is consistent and easily observable even at the walk; and a grade five is severe lameness with a reluctance or inability to bear weight on the affected limb.

Altering Work Intensity

- When managing certain chronic soundness issues, you may have to lessen the intensity of the work the horse performs.
- For example, a hunter or jumper with a chronic soundness issue may be able to stay sound for jumping if the jumps are kept low and he's asked to jump less often.
- With horse owner, veterinarian, and farrier working together, many chronic issues can be managed so that the horse stays usable and comfortable.

Switching Careers 1

- Many performance horses are retired as trail horses when they are no longer sound enough to perform their original sport.
- Trail riding can also be a demanding occupation, especially if steep hills or faster speeds are part of the workout, so intensity levels must be monitored based on the horse's comfort and abilities.
- Most horses enjoy and benefit from slow trail rides on easy terrain.

The prognosis for each horse is different depending on the diagnosis and possible treatments. Even horses with a grade zero and no apparent lameness will likely have some blemishes or issues with their legs or hooves, even if it doesn't currently cause lameness. Certain soundness issues are acute and can be treated and resolved with no long-term effects on use, while others are chronic and must be managed for the horse to stay comfortable and usable. In addition, horses' soundness and physical abilities change over time, just as with aging humans. As most horse owners who have paid for prepurchase exams (in which a vet examines a horse before purchase) know, it's hard to find a 100 percent "clean" horse.

This is where the term *serviceably sound* comes in. Serviceably sound means the horse is sound enough to comfortably perform his current uses. For example, a horse with certain leg or hoof issues may stay sound for flat work but become lame if asked to jump, or a trail horse may be steady on easy terrain but have a difficult time negotiating hills.

Switching Careers 2

- Good-tempered horses are often retired as lesson horses when soundness issues force them to leave higher levels of competition. Because these horses are well trained, they have a great deal to offer those learning the sport.
- Many chronic soundness conditions benefit from light, regular exercise, which beginner lesson programs can provide.
- Horses needing lower intensity performance often make good, serviceably sound mounts for beginning riders purchasing their first horse.

Pasture Sound

- Although some horses can be ridden well into their twenties and even thirties, other seniors or horses with soundness problems must be retired from riding altogether.
- If a horse is no longer sound enough for riding but is comfortable enough to meander around his pen, he is considered pasture sound.
- Even fully retired horses need regular attention from the veterinarian and farrier to remain pasture sound and to stay healthy.

ARTHRITIS AND JOINT DISEASE

Degenerative joint disease and osteoarthritis are likely the most common causes of lameness in horses

The terms *degenerative joint disease* (DJD), *osteoarthritis,* and *arthritis* are often used interchangeably. However, many vets classify osteoarthritis or arthritis as the advanced stage of DJD, where changes go beyond the joint capsule and cartilage to affect the bone.

A joint is formed where two bones meet and is made up of the smooth articular cartilage at the ends of the bones and surrounded by the joint capsule. Lining the joint capsule on the inside is the synovial membrane, which produces synovial fluid. All of these parts allow a normally loaded joint to move without friction or pain.

However, when a joint is injured, unevenly loaded, or sub-

Healthy Pastern and Fetlock

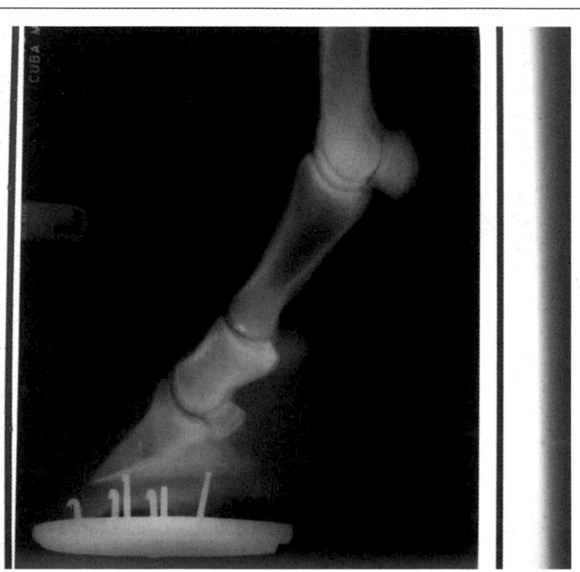

- This x-ray shows the lower leg of a horse not affected by DJD or osteoarthritis.
- Healthy joints have adequate space between bones and healthy cartilage, which allows for friction-free movement.
- This x-ray allows us to clearly visualize the bones and joints of the lower leg.
- While osteoarthritis can affect both high-motion joints like the fetlock and low-motion joints like the pastern, it's usually more painful in high-motion joints.

DJD/Osteoarthritis

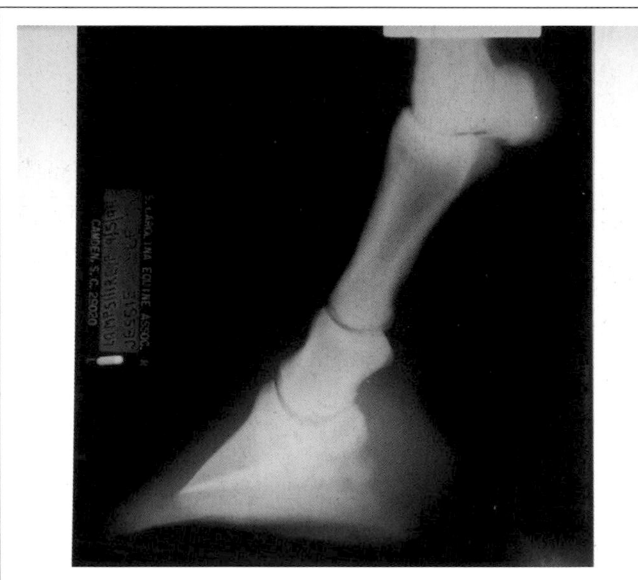

- Because osteoarthritis is considered the end stage of DJD, it can be diagnosed via x-ray.
- X-rays may how loss of cartilage, narrowing of the joints space, bone spurs and other telltale signs.
- What the x-rays show can help the farrier trim the horse, help the owner and veterinarian determine the proper workload for the horse, if any, and influence the pain management of the horse, including oral NSAIDs and possible joint injections.

ject to constant concussion due to poor conformation or athletic pursuits, a degenerative cycle can begin. This cycle starts with synovitis, which is inflammation of the synovial membrane and is often accompanied by capsulitis, an inflammation of the fibrous joint capsule. The inflammation may be visible as effusion around the joint because one of the body's responses to the inflammation is an increase in synovial fluid production. Synovitis is usually accompanied by heat and pain.

If the inflammation continues untreated, the body's cellular and chemical response ultimately has a negative impact on the joint, diminishing its shock-absorbing and lubricating abilities. Eventually this process degrades the articular cartilage, which then leads to bone changes. Concussion from use can also damage the articular cartilage. DJD is painful to the horse and decreases the mobility of the affected joints.

In young horses, joint issues are often due to osteochondrosis (see page 186). Older horses, on the other hand, are often affected by DJD largely because they have experienced more years of wear and tear on their joints.

Healthy Hock

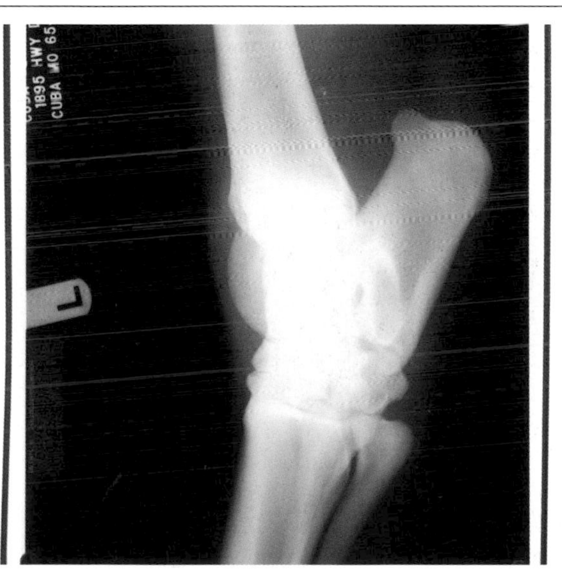

- This photo of an x-ray shows a hock not affected by osteoarthritis or DJD.
- When we put a horse in training, the bones must adapt. It takes about a year for the bones to adapt, so strenuous work must be introduced gradually over a long period of time.
- Conditioning a horse slowly and gradually, working him on forgiving surfaces, and keeping his hooves balanced can all help prevent osteoarthritis.

Hock Osteoarthritis

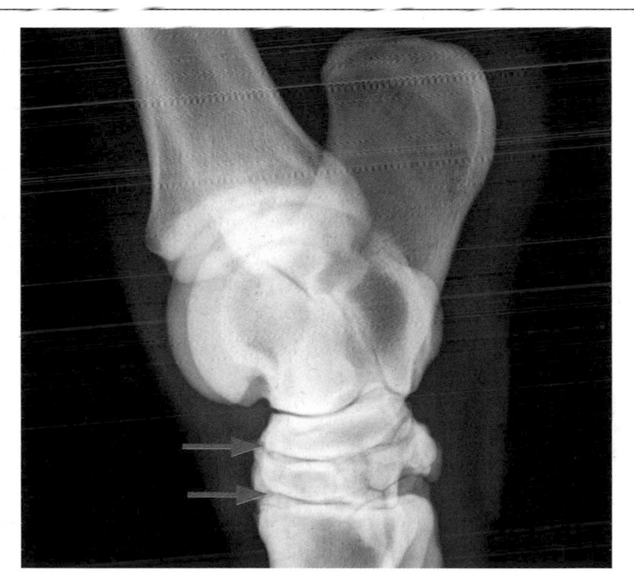

- This x-ray is of a hock affected by osteoarthritis.
- Joints affected by osteoarthritis will usually be stiff and have an impaired range of motion.
- Sometimes joints affected by osteoarthritis can make a crackling noise.
- Once articular cartilage or bony changes have occurred, treatment can only be aimed at managing DJD. Nonetheless, management is key to control further destruction.

PREVENTION OF JOINT DISEASE

Learn what contributes to joint disease and how to spot it early for a better outcome

Although it may be impossible to completely prevent degenerative joint disease (DJD) and osteoarthritis, there are contributing factors that you can help to mitigate. These factors include the management discussed on pages 90–91, such as good footing and regular farrier care. It's also important to keep your horse conditioned and at a proper weight.

Another major contributing factor to DJD is conformation. Horses with poor conformation load their joints unevenly, predisposing them to DJD. This makes buying and breeding horses with good conformation important for more than just winning horse shows.

It's also imperative to catch joint disease early, before car-

Jumping

- Different disciplines place added extension and concussion on certain joints.
- For example, jumpers may be especially prone to DJD of the hock, to which both cantering and pushing off to jump could contribute.
- Jumpers can also be prone to arthritic changes of the pastern. This isn't surprising if you watch a jumper landing, where a great deal of concussion and extension is placed on the front legs.

Reining

- Joint problems associated with the hock joints are one of the most common factors limiting performance in reining horses.
- If you imagine the moves required of a particular performance horse, you can deduce which joints are most susceptible to DJD.
- Consider the concussion and extension placed on the hock during moves such as the sliding stop and spin.
- Similarly, reining horses are prone to DJD of the stifle joint.

tilage is damaged, and before there are permanent bony changes or fibrosis of the joint capsule. The further joint disease progresses, the more permanent damage is done. The first signs of joint trouble are often a decrease in performance (see page 84). Other early signs include swelling or effusion around the joints, possibly accompanied by heat. An affected horse may also be mildly or intermittently lame and have decreased mobility in the painful joint. While it costs money to have the vet out, it will save you money in the long run to catch and treat problems early.

YELLOW LIGHT

Although a single traumatic injury or developmental issues can lead to arthritis, use trauma, associated with concussion and overextension, is the most common cause. Joint disease affects joints with a great deal of movement, like the knee and fetlock, as well as lower-motion joints, like the pastern. Certain sports or disciplines place extra strain or concussion on specific joints.

Cutting

- As with reining, cutting horses are susceptible to DJD in both the hock and stifle joints.
- Ringbone, a type of osteoarthritis in the pastern area, is also common in working Quarter Horses that perform jarring work, such as cutting.
- Conformation defects like upright pasterns can make a horse more prone to ringbone because of the added concussion.

Pulling

- Horses that pull carts or wagons are also prone to DJD in the hock joints.
- Pulling horses use their hocks to help move the load forward or up a hill.
- Horses pulling carts or wagons (other than show horses) also commonly travel on hard ground or pavement, which adds to the concussion on all joints.
- Many heavier draft horses have short, upright pasterns, which can increase the concussion placed on the pastern area and make them more susceptible to ringbone.

DIAGNOSING DJD

Getting an early and accurate diagnosis of joint disease can help halt its progression

If you and your vet suspect a joint problem, the vet will conduct a soundness exam, checking the horse for changes in movement. The vet will also inspect the joints for heat, swelling, and decreased range of motion. Keep in mind that pain in one area can stem from pain in another area. For example, joint pain in the lower leg could make the horse's hip sore as he changes the way he moves to compensate. Therefore, the veterinarian will conduct a thorough exam, often including nerve blocks to isolate the primary source of pain.

A joint block may also be performed, which numbs the joint. Similar to a nerve block, if the horse is suddenly sound after the block, then that joint is likely the cause of his pain.

Changes in Performance

- It's important to know your horse well. Know what is normal behavior and what is standard performance for him.
- When a horse begins behaving differently under saddle, or his performance starts to slide, he's telling you something.
- Poor performance or changes in movement are usually the first signs of joint trouble (and other soundness problems).
- If joint problems are caught early, permanent damage may be avoided by altering training and implementing a treatment program.

Swollen Joints

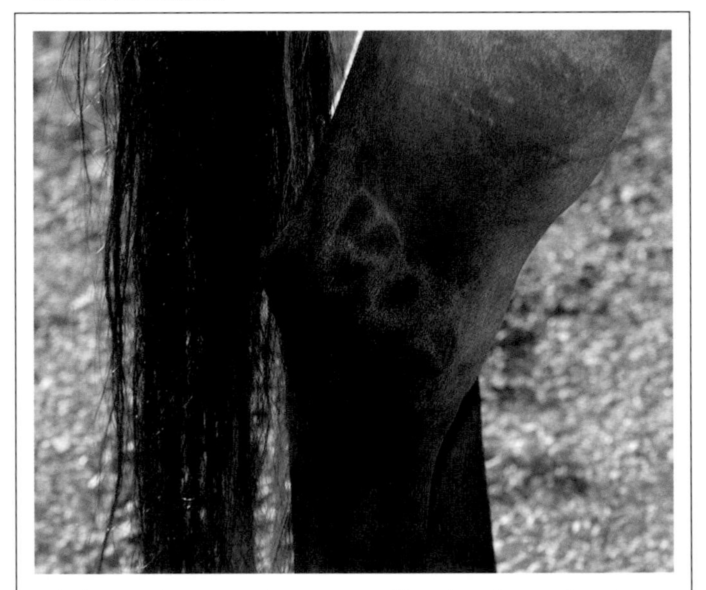

- Inflammation around a joint, or a joint area distended by excess fluid (effusion), is a possible sign of joint trouble.
- Joints with a greater range of motion tend to have more visible swelling and fluid distension than joints with less movement.
- Whether the swelling is accompanied by heat or not, consult your veterinarian.
- Stopping joint disease at this stage may mean you can save your horse from cartilage and bone damage.

Performing a joint block may also allow your vet to examine a bit of the synovial fluid at the same time. The composition of the fluid is an indicator of joint health. Synovial fluid analysis is helpful in diagnosing degenerative joint disease (DJD) and indicating the level of degeneration.

X-rays are another common tool for diagnosing DJD and osteoarthritis but will not show soft tissue changes or the beginning stages of joint disease. However, x-rays can show loss or narrowing of the joints space, extensive cartilage damage, calcification/ossification, and bone changes caused by osteoarthritis. X-rays can also show other bone problems in the area, such as fractures or bone chips, which could be primary or secondary problems.

Although MRIs are more expensive and less widely available than x-rays, the resulting images are excellent for diagnosing joint problems. Ultrasound can also show soft tissue abnormalities that x-rays cannot. Arthroscopy, in which a tiny arthroscope is used to examine the joint, is extremely useful in determining the degree of cartilage degradation, even before any changes are visible via x-ray.

Synovial Fluid Analysis

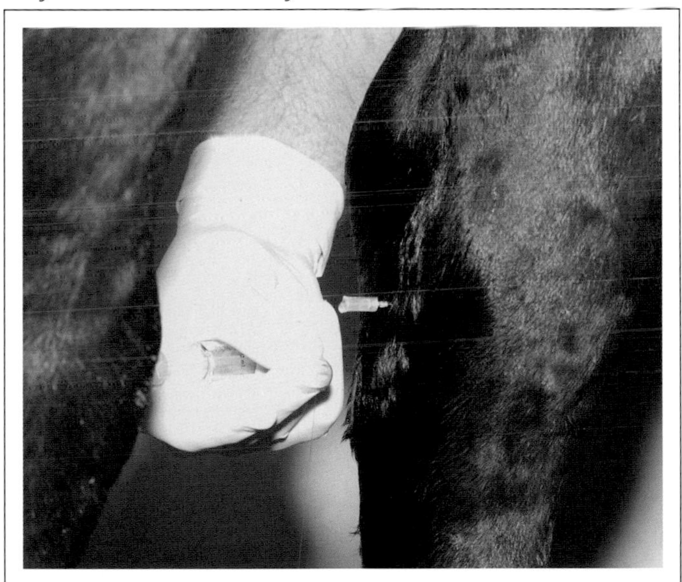

- The vet can use a needle to take a sample of the synovial fluid in the horse's joint and analyze its consistency and composition for signs of trouble.
- Synovial fluid analysis is a great tool for early diagnosis of degenerative joint disease, as it can indicate a problem before bone or extensive cartilage changes occur.
- Synovial fluid analysis also allows the vet to rule out the serious but less common infective or septic arthritis.

Thermography

- Infrared thermography measures heat emitted from the body and produces images showing the heat in colors.
- Thermography can be a helpful tool in diagnosing where a problem is located by comparing one leg to the other.
- Hip and shoulder joints can't be easily manipulated or isolated via blocks but can be imaged using thermography.
- Thermography shows changes in blood flow and hot/cold spots; it can indicate potential problem areas before other methods.

TREATMENTS FOR JOINT DISEASE

Most treatments for joint disease help slow the cycle of degeneration and keep the horse comfortable

There are many treatment options for degenerative joint disease (DJD), depending on the stage of degeneration. Treatment options are influenced by what activities you'd like the horse to continue to perform and how much you can afford to spend on treatment.

For horses just starting to show signs of joint trauma and horses with synovitis (inflammation of the synovial membrane in the joint), a period of rest is often prescribed to let the joint recover and to reduce inflammation. Rest is usually combined with light hand walking and cold therapy, such as ice wraps applied to the affected joint periodically for several days. Your vet may also recommend treatment with DMSO

Intra-articular Injections

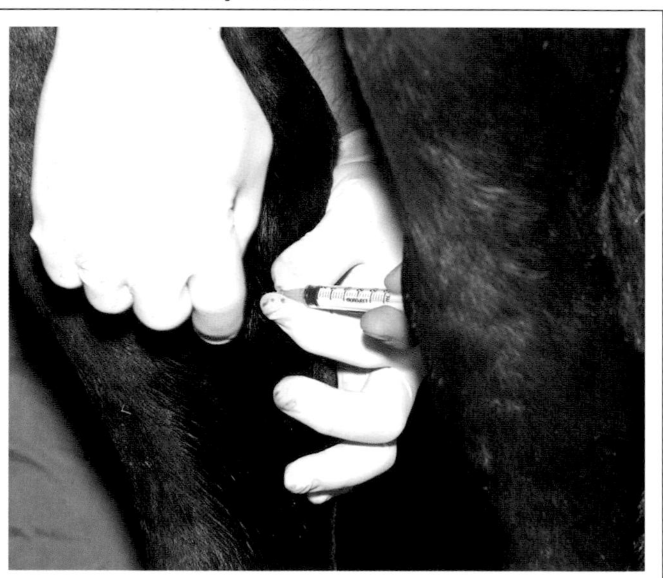

- Intra-articular injections are injected into the joint itself.
- Corticosteroids and hyaluronic acid (HA) or a combination of the two are the most common intra-articular injections.
- By working as anti-inflammatories, these injections help slow the damaging cycle caused by joint inflammation.
- Hyaluronic acid occurs naturally in the joint, and injections of HA may help lubricate the joint and stimulate natural HA production.

Intramuscular or Intravenous Injections

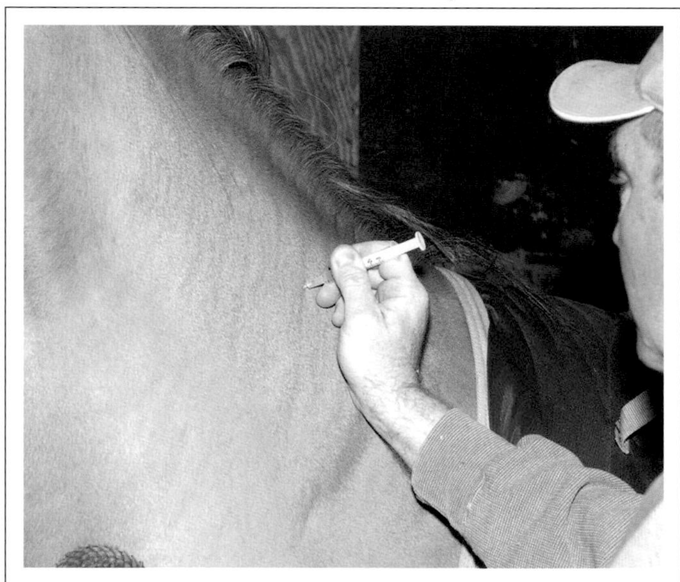

- Intramuscular injections are injected into the muscle, while intravenous injections are injected into a vein.
- Polysulfated glycosaminoglycan (PSGAG) such as Adequan can be injected into the muscle or the joint.
- PSGAG helps inhibit enzymes that break down joint cartilage and fluid, and stimulate healing and natural hyaluronic acid production.
- Hyaluronic acid, mentioned previously as an intra-articular injection, can also be given intravenously.

(see page 196) and possibly wrapping the affected joint. Other physical therapies can also be beneficial.

Since inflammation is the main cause of the destructive cycle within the joint, most treatments for DJD are aimed at reducing inflammation. Nonsteroidal anti-inflammatory drugs (NSAID), such as phenylbutazone, commonly known as bute, help with pain and inflammation. Joint injections are one of the most common treatments for DJD. (See also page 102 for more information.) Other treatments will be reviewed in Chapter 18.

In some cases, joint lavage is helpful. It is a process where the joint is essentially flushed or washed out. In horses with active synovitis, joint lavage may help remove toxins and damaging inflammatory substances as well as accumulated tissue debris. Joint lavage, along with antibiotics, is used on horses with septic (infective) arthritis, where a bacteria or virus has invaded the joint.

If bone fragments are present, leading to joint inflammation, surgery to remove them may be recommended. Joint problems can also cause or be caused by bone fractures.

New NSAID

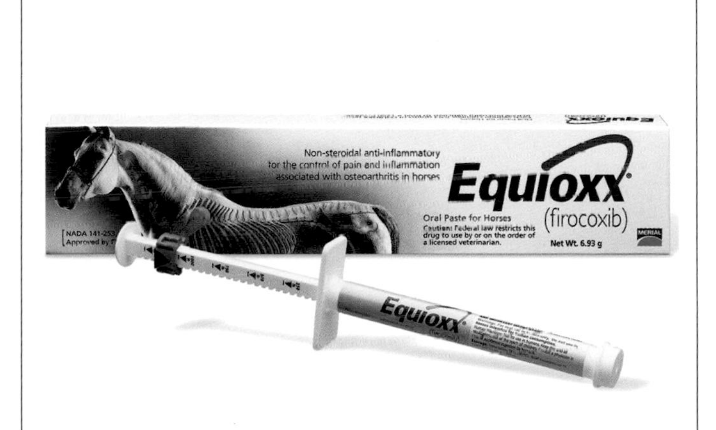

- The COX-1 sparing inhibitor firocoxib (brand name EQUIOXX) is an NSAID like bute, but EQUIOXX highly selects for COX-2 enzymes and spares COX-1 enzymes, which help in protecting the stomach lining and is therefore associated with fewer gastrointestinal side effects compared to bute.
- In field trials, EQUIOXX was shown to help relieve the pain and inflammation associated with osteoarthritis.
- EQUIOXX, a perscription drug, is currently available in paste form, which provides relief for twenty-four hours.

Joint Fusion

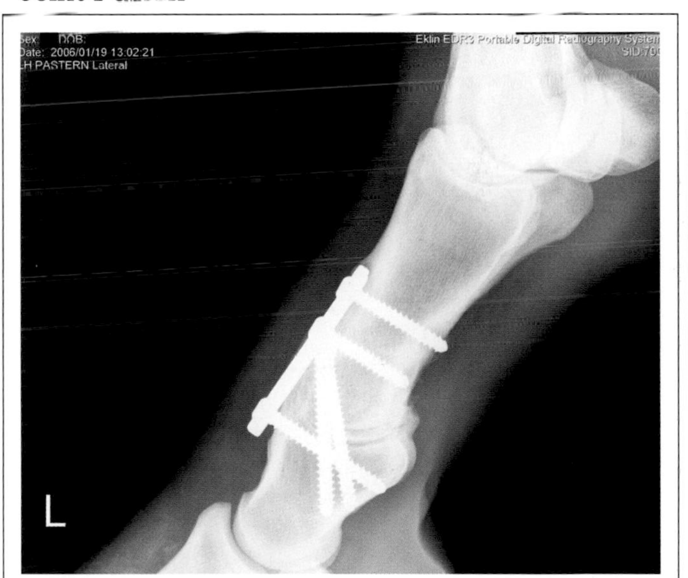

- When cartilage degeneraes, one bone may eventually come in contact with the bone on the other side of the joint, and in low-motion joints they may fuse.
- Arthrodesis is the surgical fusing of two bones within a low-motion joint.
- This horse suffered an injury that led to septic pastern joint and osteomyelitis. Pastern arthrodesis was performed, and the horse is now sound as a trail horse.
- In end-stage osteoarthritis, arthrodesis can help eliminate joint pain and stabilize the joint.

JOINT SUPPLEMENTS

There are many joint supplements available over the counter with natural ingredients that may be beneficial

Oral joint supplements can be purchased via the Internet or at virtually any tack and feed store. The three most common ingredients in oral joint supplements are glucosamine, chondroitin sulfate, and methylsulfonylmethane (MSM). Both glucosamine and chondroitin sulfate are naturally part of articular cartilage. The theory behind feeding them as supplements is that they may help protect cartilage or reduce inflammation and pain. MSM is a sulfur supplement that has the same aims.

These ingredients are also popular in human joint supplements, but the effectiveness is still being studied. Extensive studies in horses have not been conducted either, and because the horse's intestinal tract is quite different than ours,

Joint Supplements

- There are endless choices when it comes to joint supplements.
- Joint supplements contain varying amounts and combinations of ingredients.
- The quality of ingredients can vary, and likely affects how well your horse's body can absorb or utilize the supplement.
- If you choose to feed your horse a joint supplement, ask your vet for a recommendation.

Ingredients

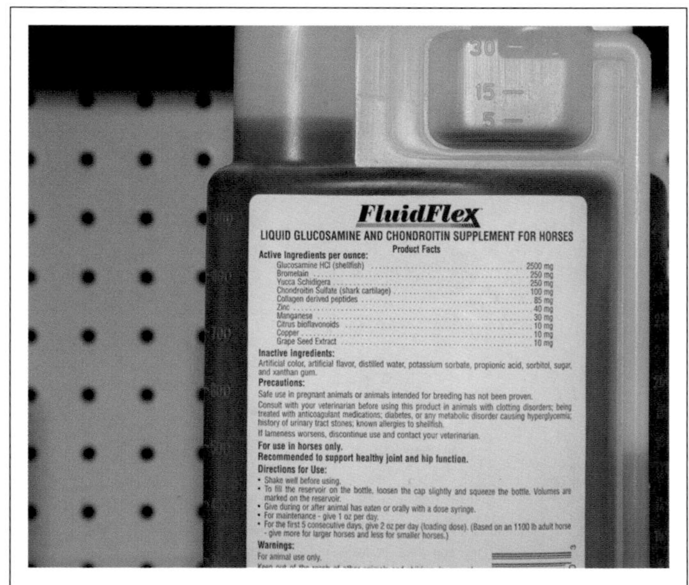

- Oral supplements get their glucosamine and chondroitin from various sources, including cow tracheas, sea mussels, and shark cartilage.
- There are also two types of glucosamine: glucosamine sulfate and glucosamine hydrochloride. Which types are best for horses has not been extensively researched.
- Products contain varying amounts of glucosamine and chondroitin, so make sure to compare amounts of active ingredients.

it's also questionable how well horses can absorb oral supplements. Hyaluronic acid (HA) is also now available in oral forms, but its efficacy is not as well documented as intra-articular and intravenous HA injections.

Adding to the uncertainty over oral joint supplements for horses is the fact that the Food and Drug Administration (FDA) does not monitor nutraceuticals for safety or quality since they are considered food supplements, not drugs. University studies have shown that many oral joint supplements for horses do not contain the amount of the ingredients they claim. However, many manufacturers have joined the National Animal Supplement Council (NASC), which is the industry's attempt to self-police. Among other rules for compliance, NASC member products must contain what their labels state. In addition to glucosamine, chondroitin sulfate, and MSM, joint supplements may contain additional ingredients. Manganese is a component of cartilage and may work beneficially with glucosamine. Devil's claw is touted as a pain reliever and yucca as an anti-inflammatory. However, these claims are not well proven in horses.

Feeding

- Oral joint supplements come in powder or liquid forms and must be carefully measured.
- Most oral supplements are designed to be mixed in daily with grain or sweet feed, and most have a higher-dose loading phase.
- Horses generally don't mind the taste of oral joint supplements.
- To help ensure the joint supplement you buy actually has the ingredients it claims, look for a guaranteed analysis claim.

Effect on Performance

- It's unlikely you'll see a dramatic change in your horse's performance after you begin feeding oral joint supplements, but some owners do report improvements or note their horses respond better to one supplement versus another.
- Many owners feed supplements hoping they will make their horse feel more comfortable or slow further damage to the joints.
- Some horse owners provide oral joint supplements as a preventative measure to ward off DJD, but this hasn't been proven as an effective preventative measure.

EXERCISE AND JOINT DISEASE

For a horse with joint disease, exercise can be his best friend or his worst enemy

Considering use trauma is the main cause of degenerative joint disease (DJD), it may seem that a horse with DJD will be better off retired. However, this is not entirely true. While a horse with DJD may need a less strenuous career, light, regular exercise is often critical to his long-term health and comfort.

For one thing, exercise helps keep the horse at a healthy weight preventing extra strain on his joints, not to mention their hooves, muscles, tendons, and ligaments. Exercise keeps the muscles from atrophying too. Muscles help support the horse's joints, so muscles that fatigue too easily place greater strain on the joints, tendons, and ligaments. Exercise also

Trimming and Shoeing

- Proper, regular trimming and shoeing is extremely important for horses with DJD and osteoarthritis.
- Ease of breakover is one key factor your farrier should consider when trimming or shoeing a horse with joint problems.
- Added traction, such as rim shoes or studs, places additional strain on joints.
- Changes to how your horse is trimmed or shod may be necessary to help keep him comfortable if he is diagnosed with DJD.

Lifestyle

- In the wild, horses graze and move about most of the day.
- Living in a paddock or pasture, or receiving daily turnout, is helpful to horses with DJD.
- If you plan on retiring your horse with DJD or osteoarthritis, then choosing a large paddock or pasture is your best option.
- With room to move, your horse can walk around and avoid getting stiff.

improves circulation and circulates the joint fluid within the joint to help remove waste and provide nutrients to the articular cartilage.

That said, the type of riding you do may need to be altered if your horse suffers from DJD. If your horse performs strenuous work, such as jumping, cutting, endurance, and so on, his workload may need to be decreased, or a change in career may be needed depending on the severity and stage of DJD. Your veterinarian can help you determine the best exercise and maintenance plan for your particular horse.

GREEN ● LIGHT

Passive motion exercises involve the handler gently flexing a horse's joints from the ground. Stretching exercises are when the handler helps the horse stretch. Passive motion exercises and stretching can be beneficial to humans suffering from arthritis, and they likely also benefit horses if done correctly.

Strenuous Exercise

- Most high performance horses, including jumpers, dressage horses, reiners, and barrel racers, eventually retire or change careers, and DJD is the most common cause.
- An aggressive treatment strategy and careful maintenance can help keep a horse performing longer, but it's important to monitor the horse's comfort level.
- Also keep in mind that painkillers, such as bute (phenylbutazone), mask a horse's pain and allow him to perform in ways that may cause additional damage.

Light Exercise

- Light, regular exercise on good footing is the best option for horses with DJD.
- Regular exercise means three times a week or more.
- Irregular, hard exercise is not advisable for any horse and especially problematic for horses with DJD.
- A balanced rider of a suitable size is also important for horses with joint problems, as excessively heavy or unbalanced riders add strain to the joints.

RINGBONE AND SIDEBONE

Ringbone and sidebone are types of osteoarthritis that affect the lower leg with varying impacts on soundness

Ringbone gets its name because of the bony ridges it forms around the pastern bones. Sidebone, in turn, creates bony growth along the sides of the pastern area. These bony changes are the result of osteoarthritis (see page 80 for symptoms, diagnosis, and treatment of joint disease and osteoarthritis).

Horses at risk for ringbone and sidebone often have been worked on hard surfaces for many years, placing a great deal of concussion on the lower front legs. Poor shoeing or conformational defects, including upright pasterns, long toe/low heel, and toeing in or out, can also predispose a horse due to uneven loading of the joint.

Healthy Pastern and Coffin Joints

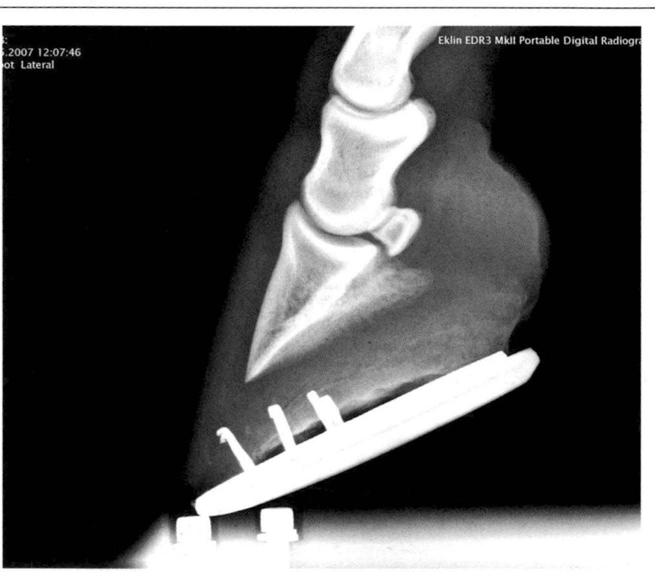

- This x-ray shows a healthy pastern and coffin joint without the arthritic changes associated with ringbone or sidebone.
- Notice there are no abnormal bone formations (calcification/ossification) around the pastern bones.
- Some horses eventually develop both high ringbone and low ringbone together.
- Ringbone and sidebone can be diagnosed using x-ray due to the bony changes they create.

Low Ringbone

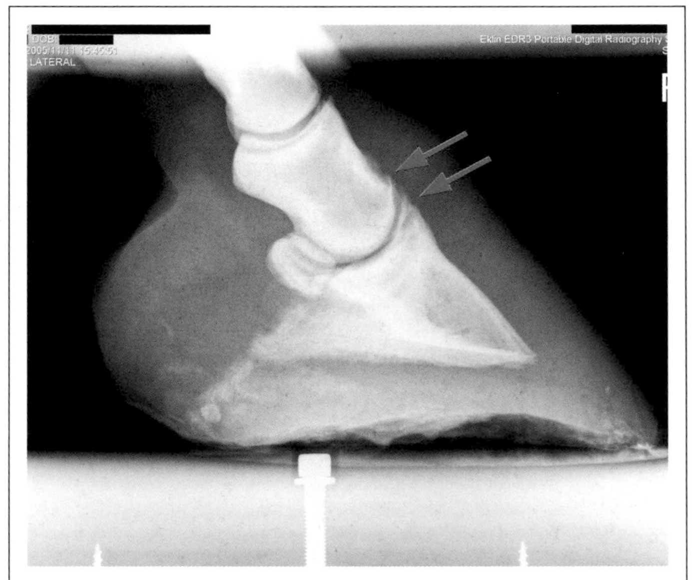

- This x-ray is of a nineteen-year-old horse with low ringbone causing chronic lameness.
- Although less dramatic than the x-ray of high ringbone, you can see the changes where the short pastern and coffin bone meet.
- Low ringbone often causes more problems than high ringbone due to its location near a higher-motion joint.
- Ringbone usually affects both front legs, although one leg may be worse than the other.

As with other types of osteoarthritis or degenerative joint disease (DJD), ringbone starts out as inflammation in the joint area, often accompanied by heat, pain, and swelling. Eventually bony proliferations (calcification/ossification) form around the pastern bones.

High ringbone is calcification where the long and short pastern bones meet, and low ringbone is calcification where the short pastern and coffin bone meet (which you may be able to feel in the coronary band area). Because the joint where the short pastern bone and coffin bone meet is a high-motion joint, low ringbone is often more painful to the horse than high ringbone. Ringbone can be further divided into articular and nonarticular, with nonarticular at the edge of the joint and articular involving the joint surfaces and proving more painful to the horse.

Sidebone is calcification of the lateral cartilages that flank the coffin bone. You can feel advanced sidebone at the sides of the hoof above the coronet, and it may be sore to the touch. While sidebone can affect hoof expansion, horses with sidebone often remain sound and can continue working.

High Ringbone

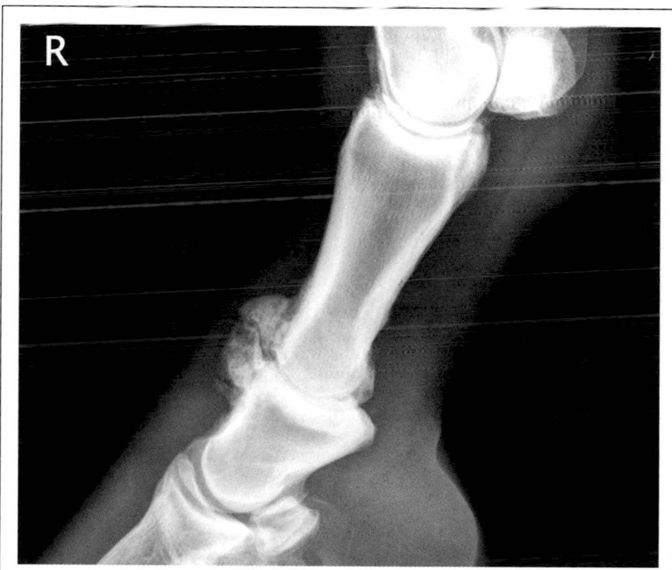

- This x-ray is of an eighteen-year-old horse with high ringbone that caused lameness.
- You can clearly see the boney formations at the front of the leg where the long pastern bone and short pastern bone meet.
- In high ringbone the joint might eventually fuse since it's a low-motion joint, or, in some cases, it can be surgically fused for stabilization.
- With treatment, such as IRAP injections and non-steroid anti-inflammatory medications, the horse is comfortable for trail riding.

Sidebone

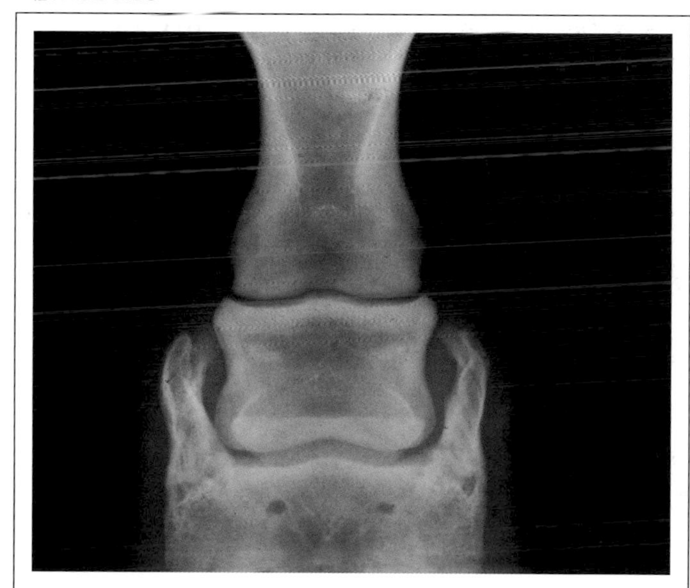

- This x-ray is of a twenty-year-old horse with sidebone. The horse is sound and is used as a jumping lesson horse.
- You can easily see the boney formations along each side of the hoof, rising up toward the coronet area.
- For horses with ringbone and sidebone, it's important for the farrier to work to balance the hooves.
- Easing breakover or moving breakover back may also be recommended for horses with ringbone or sidebone.

SPAVIN AND SPURS

Bone spavin, bog spavin, and bone spurs can all affect the hock as a result of osteoarthritis

The horse's hock, or tarsus, is made up of six bones and four joints. The large, upper joint, known as the tibiotarsal joint, is responsible for the majority of movement in the hock. The other three joints are low-motion joints, meaning the bones they connect don't move much when the horse bends his leg.

Bog spavin affects the high-motion tibiotarsal joint and presents as excess fluid around the hock. Bog spavin is usually associated with synovitis and degenerative joint disease (DJD) caused by use trauma.

The swelling or effusion often appears suddenly but usually becomes chronic if the underlying cause is not addressed. If

Symptoms

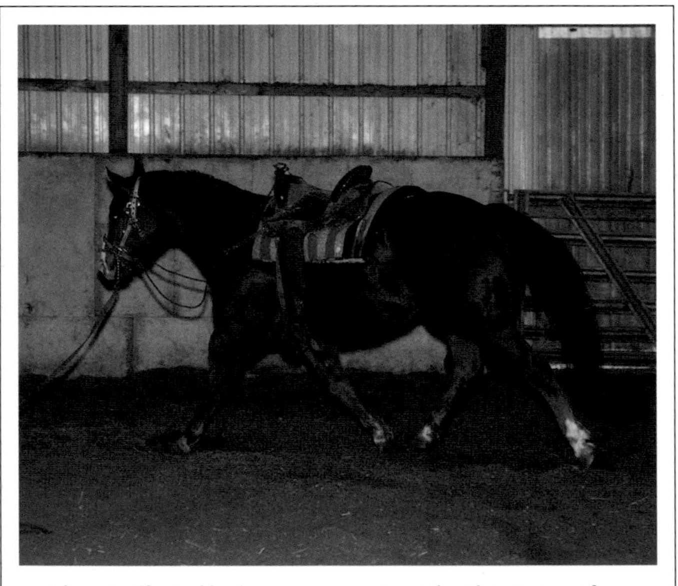

- A horse affected by bone spavin, or hock arthritis, often travels stiffly in the hind end, stabbing his hind legs as he trots.
- Bone spavin is usually present in both hind legs, but lameness may be visible in one leg more than the other.
- As with other types of arthritis, a horse with bone spavin may warm up and work out of his lameness.

Joint Injections

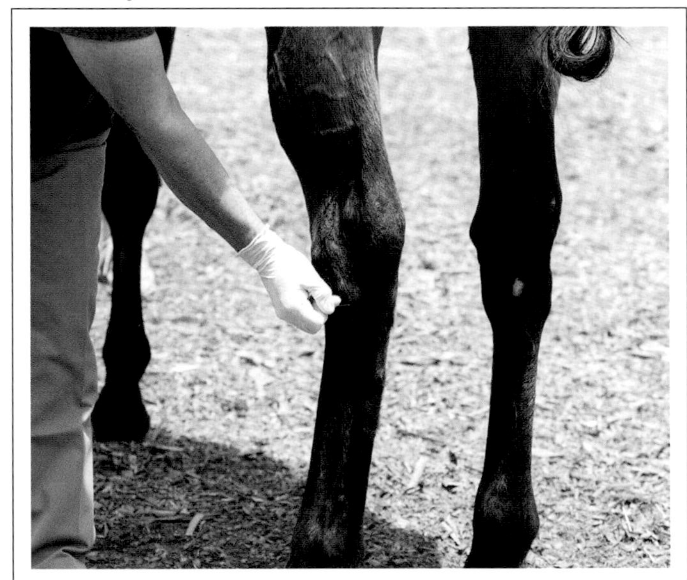

- Joint injections are the most common treatment of joint disease and may help relieve inflammation and improve joint health.
- Joint injections are often recommended for horses with bog spavin or horses with bone spavin.
- The prognosis for horses with bone spavin depends on how aggressive the treatment is and what level of work the horse is expected to perform.

bog spavin is treated before DJD affects the articular cartilage or causes bone changes, then the prognosis is often good.

Bone spavin (also called distal tarsal osteoarthritis or distal tarsitis) is likely the most common cause of hock lameness. Bone spavin is DJD of the hock, usually affecting two small, low-motion joints in the lower hock.

Bone spavin is usually associated with use trauma. Uneven loading of these joints due to the type of activity the horse performs or his conformation can increase the likelihood of bone spavin.

ZOOM

Bone spurs, also known as osteophytes, are sharp bone growths at joint margins that can occur in the hocks or elsewhere as a result of osteoarthritis. When articular cartilage degenerates due to joint disease, bone spurs may form as the horse's body attempts to stabilize the joint. Bone spurs are usually visible via x-ray.

Bog Spavin

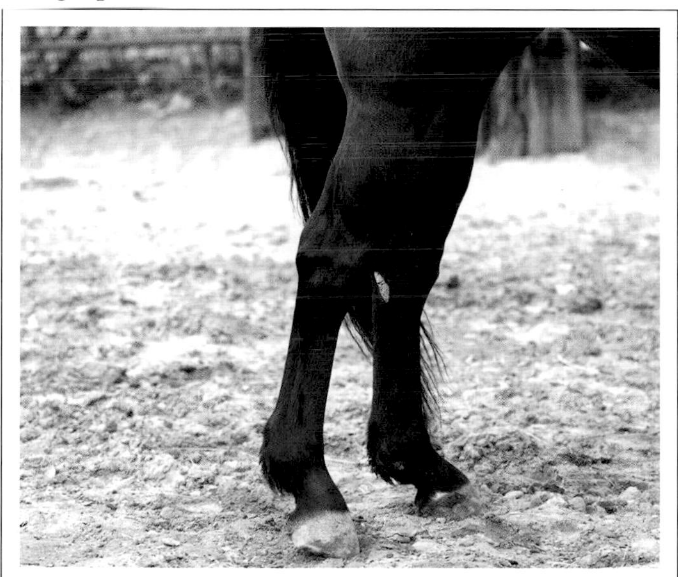

- Bog spavin can be easily spotted, as excess fluid will accumulate around the sides and front of the hock.
- At first bog spavin may not be accompanied by any lameness.
- Since horses with bog spavin are often sound, some owners consider it a blemish and have the fluid drained.
- Draining the fluid associated with bog spavin is only treating the symptom and not the cause, so the swelling will likely return.

Bone Spurs

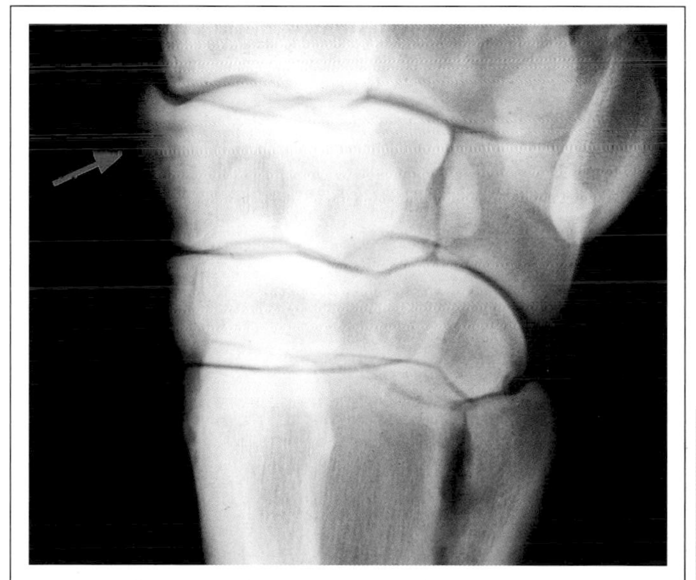

- Bone spurs (osteophytes) often occur in the hocks and knees and other joints.
- This x-ray is of a bone spur in the carpus.
- Bone spurs are one possible result of DJD and osteoarthritis and are the body's attempt to help stabilize a joint.
- How much pain bone spurs cause or how much they limit the joint's mobility depends on their size, location, and which joint(s) they affect.

OSSELETS

Osselets cause swelling and often bony changes to the front of the fetlock joint

Osselets generally refer to degenerative joint disease (DJD) of the fetlock joint. As with DJD in other areas, osselets start with synovitis and capsulitis (inflammation of the synovial membrane and joint capsule), causing swelling at the front of the fetlock and possibly at the sides. Swelling at the sides of the fetlock is commonly called windpuffs. At this stage, osselets are called "green osselets," and there is usually heat, swelling, and pain in the area. If caught at this stage, osselets can often be treated with anti-inflammatories, rest, and cold therapy. Successful early treatment may prevent DJD and bony arthritic changes.

If the problem is not treated early, bony growths will form

Training

- Young horses put in hard training may develop osselets in part because their muscles fatigue, not able to help support the joint.
- Osselets are associated with hyperextension of the fetlock joint, where the joint drops too far toward the ground at high speeds.
- Former racehorses often have osselets, which may or may not affect their second careers.
- When osselets are forming, the horse may travel with a short choppy gait due to the pain.

Flexion

- Green osselets are associated with pain, heat, and swelling at the front of the fetlock joint that may lead to joint disease or osteoarthritis.
- Horses with osselets often experience pain when the joint is flexed and/or a reduction in range of motion for the joint.
- The bony proliferations associated with formed osselets can be felt along the front of the fetlock, where the cannon bone and long pastern bone meet.

along the lower front end of the cannon bone and the upper front end of the long pastern bone. As with ringbone, how much pain osselets cause the horse long term depends on how much they interfere with the joint—the articular cartilage and the joint capsule. The bony changes associated with osselets also sometimes interfere with the point of attachment of the digital extensor tendon at the front of the leg, causing pain and lameness.

Osselets usually affect both front legs and are a common problem for racehorses that often experience hyperextension of the fetlock joint and receive hard training at a young age. Horses with short, upright pasterns, who are worked on hard ground or are subject to poor trimming/shoeing are also at risk for developing osselets.

As with other types of DJD, a horse with osselets may travel with a short, choppy gait or point the foot most affected by the osselets. Bone chips in the area can present similar symptoms as osselets and may need to be removed. See page 80 for complete symptoms, diagnosis, and treatment of joint disease and osteoarthritis.

Swelling

- A green osselet usually refers to synovitis and capsulitis of the fetlock joint presenting as swelling at the front and sometimes the sides of the joint.
- Green osselets only affect the soft tissues and therefore cannot be viewed via X-ray.
- If you notice windpuffs or joint swelling around the fetlock or other joints, call your veterinarian so that the cause can be addressed before it progresses into chronic synovitis or leads to DJD.

Osselets

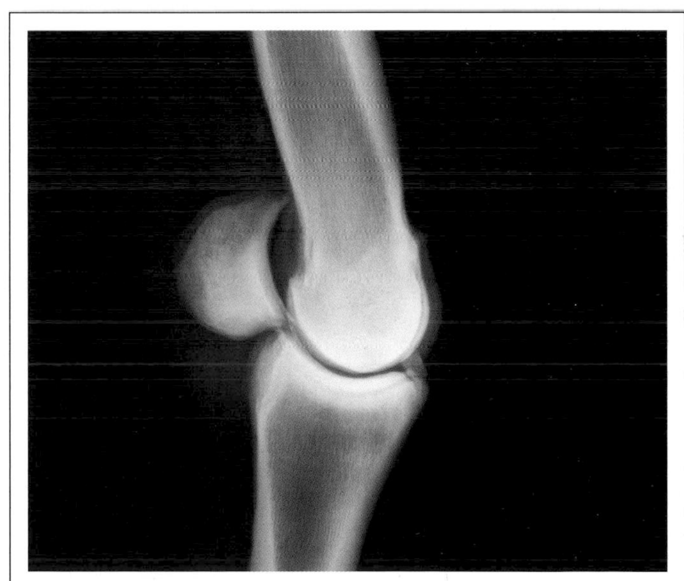

- This x-ray shows a chronic fetlock osselet on a seven-year-old Thoroughbred racehorse.
- The osselet was treated with surgical debridement of the fetlock joint and joint lavage.
- The horse retired from racing and is now sound as a dressage horse.
- As with other types of osteoarthritis, osselets may be partially prevented by slow and gradual training that allows the bones, tendons, and ligaments to adapt over time.

LUXATIONS

A dislocation can occur in virtually any leg joint with serious consequences

Just as humans can experience dislocations in their arms or legs, horses also can experience dislocations in their legs. Full dislocations are medically called luxations, and partial dislocations are called subluxations. Luxations take place when the soft tissue structures that support the joint are torn, stretched, or cut. Soft tissues supporting the joint include the joint capsule, tendons, ligaments, and muscles. The soft tissue damage leading up to the luxation can take place slowly, such as muscle atrophy, or it can occur traumatically in the form of an injury.

Horses that experience a full luxation usually cannot bear weight on the limb. The rider or handler may hear a "pop"

Pastern Luxation 1

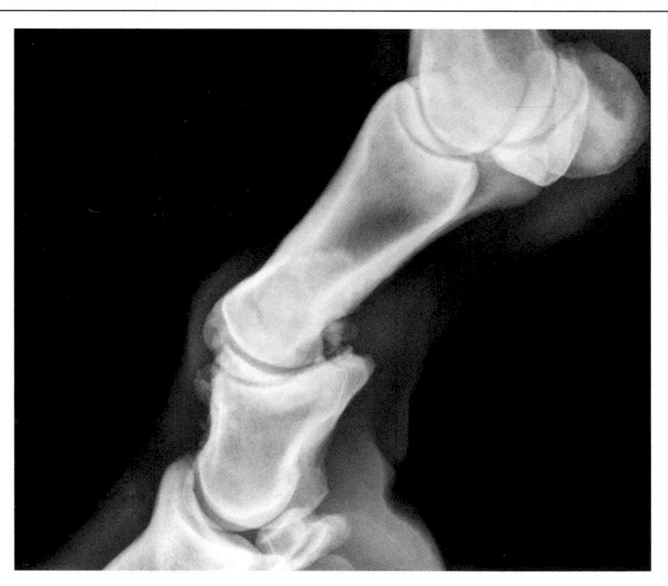

- This x-ray shows a dislocation of the pastern joint. The long pastern bone has slid forward from the short pastern bone. Unfortunately due to severe lameness and a poor prognosis, this horse was euthanized.
- For pastern luxations that are treatable, long term casting is usually part of the treatment.
- Surgical arthrodesis (fusing) is also often used in cases of pastern luxation.
- Whatever the treatment, osteoarthritis usually develops as a result of a pastern luxation.

Pastern Luxation 2

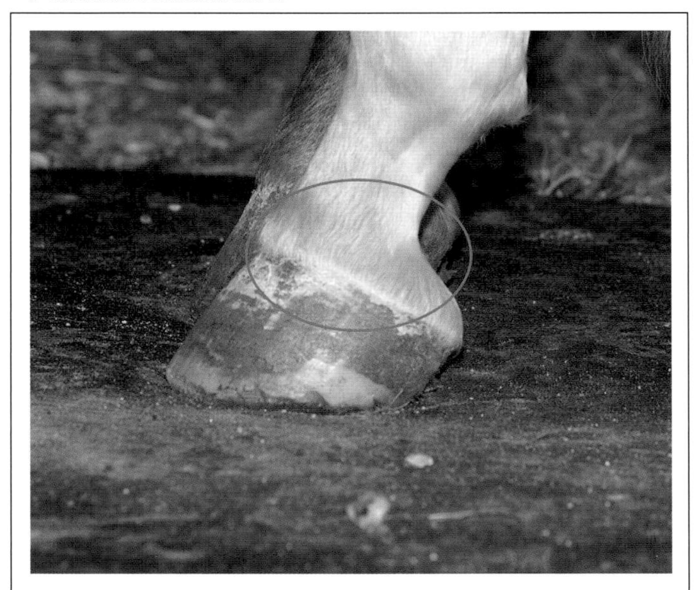

- Luxations are most common in the hock, fetlock, and pastern.
- Luxations to the pastern joint may happen in combination with a fracture to the short pastern bone and are usually caused by injury.
- Pastern luxations may appear as open or closed injuries (noninjured pastern shown in the photograph above).
- Lameness varies depending on the extent of the injury but may be non-weight-bearing.

when the luxation occurs. The symptoms are similar to signs of a fracture. Horses with subluxations may be able to bear some weight on the limb. The leg affected by a luxation or subluxation may seem misaligned or specific bones may appear out of place.

If any of these signs are present, make an emergency call to the vet, and try to keep your horse still until the vet arrives. The vet will use physical examination and x-rays to determine if the horse is suffering from a luxation, fracture, or other problem.

The prognosis for luxations depends largely on where the luxation occurred and whether it's full or partial. As with humans, relocation of a dislocated joint is painful, so the horse may be anesthetized for the procedure. Luxations in the upper leg are especially difficult to treat. If the luxation is treatable, the affected part of the leg will likely be splinted or placed in a cast during healing. Some joints can be surgically fused.

Osteoarthritis eventually develops in most joints affected by luxations, and horses often cannot return to their previous athletic pursuits though they may still be rideable.

Hock Luxations

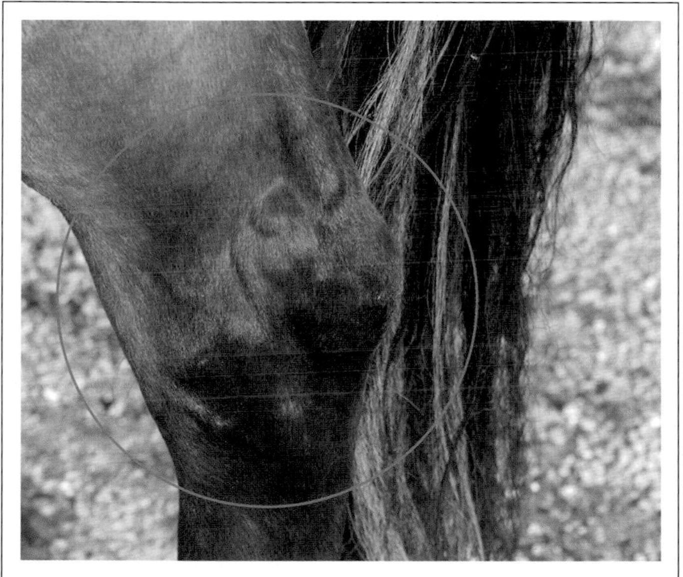

- Luxations (dislocations) occur in any of the four hock joints (nondislocated hock pictured above).
- Most hock luxations are the result of injuries—the horse slipping and falling, or getting caught up in a fence or kicked by another horse.
- The severity of a hock luxation depends on which joint is affected and can be confirmed via x-ray.
- The most severe luxation of the hock occurs in the tarsocrural joint, and in these cases the tibia may be displaced.

Fetlock Luxation

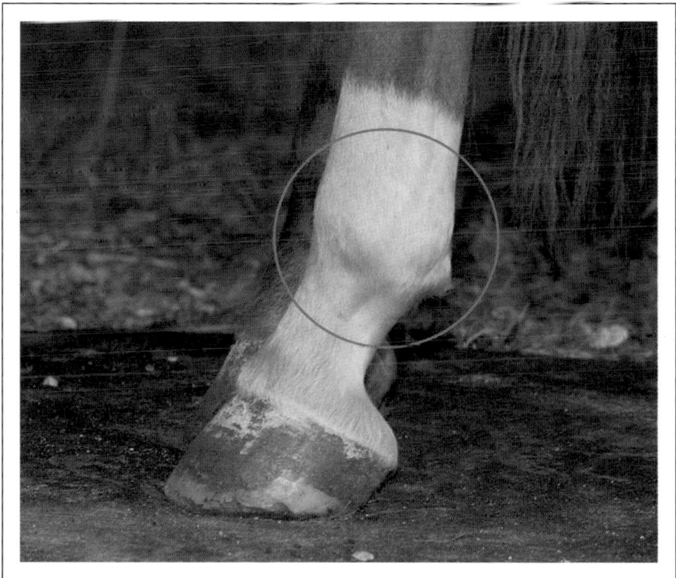

- A dislocated fetlock joint appears swollen or deformed in contrast to the normal fetlock shown in the photograph above.
- Fetlock luxations most commonly occur when the horse gets his lower leg stuck.
- Luxations usually involve damage to the ligaments, joint capsule, and nearby tendons.
- In cases of fetlock luxation, the lateral or medial collateral ligament may be ruptured.

SPLINTS AND BUCKED SHINS

Splints and bucked shins are most common in horses beginning their training

Along the sides of each cannon bone are two splint bones (also known as the second and fourth metacarpal bones in front, and second and fourth metatarsal bones in the back). Splint bones widen just below the knee and taper down, ending in the lower one third of the cannon bone.

Splint bones connect to the cannon bone via the interosseous ligament, which is elastic in young horses and usually hardens or ossifies in older horses. Splints are ossified bony lumps (calcium deposits) that form at the point of trauma. That trauma can come from the outside, such as a hoof hitting the splint or cannon bone and causing inflammation in the tissues covering the bone, or the trauma can

Splint

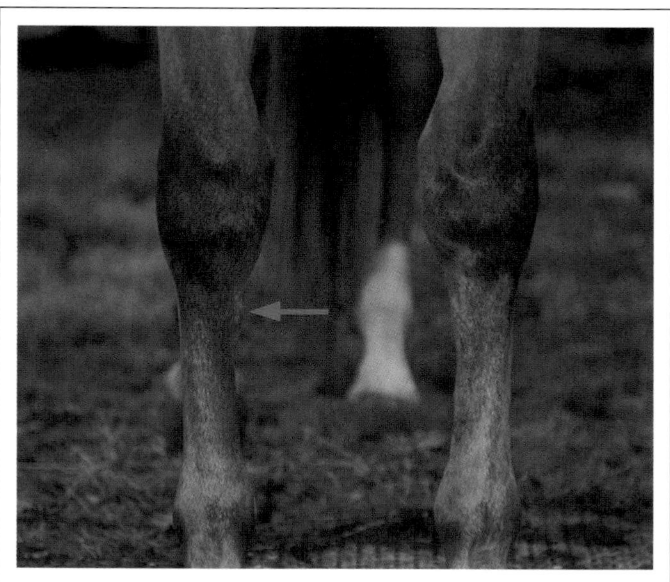

- After a splint forms, the ossified lump can likely be seen and felt but doesn't usually cause the horse long-term problems. The lump may flatten somewhat over time.
- The medial (inside) splint bone is more likely to suffer from splints because it bears more weight than the outside splint bone and is also more vulnerable to interference injuries.
- Splints most often occur on the front legs since they carry most of the horse's weight.

Splint X-Ray

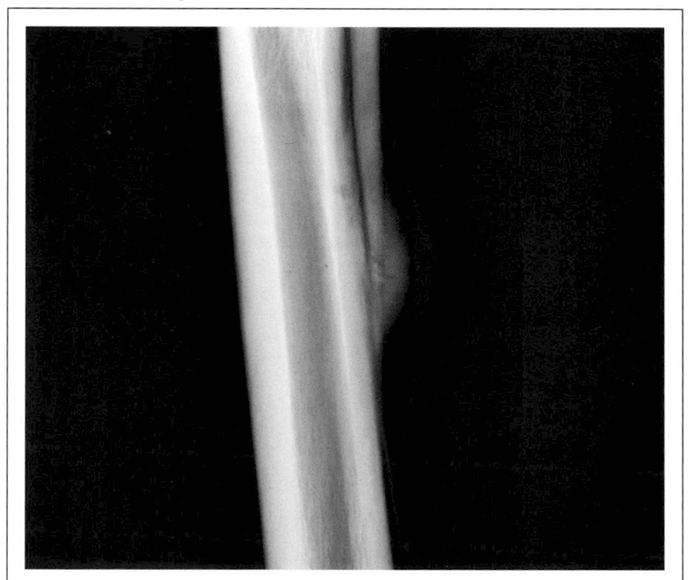

- This x-ray shows a splint bone fracture callus, which can be clearly seen on the right-hand side.
- Formed (ossified) splints can be seen on x-ray.
- X-rays can also be used when a splint is forming. If the area is highly painful, the vet may recommend an x-ray to rule out a fracture of the splint bone.
- Protective legwear, such as splint boots or sports boots can help prevent splints during riding or lunging.

come from concussion to the bone or tearing of the interosseous ligament during use.

When a splint is forming, there is usually swelling and pain in the area, and the horse will be mildly lame. A month of rest or lightened work is usually recommended and is often combined with anti-inflammatories, support wraps, and ice therapy. When the splint is no longer painful, the horse can usually return to work. Occasionally, splints cause long-term soundness issues, such as when they interfere with the knee joint or the suspensory ligament.

Bucked shins are another problem in the cannon bone area and are most common in young Thoroughbred racehorses but can affect any horse in hard training. The cannon bone adapts to the stress of training by laying down extra bone on its front surface. Inflammation results when training progresses faster than the bone can adapt. During this process, there will likely be heat and pain in the area, which may be accompanied by mild lameness or swelling. A period of rest, hand walking, icing, and anti-inflammatories is usually prescribed, after which the horse can return to training.

Bucked Shins

- When bucked shins are forming, the area will be painful and may cause discomfort during exercise as evidenced by a short-strided gait.
- Careful, progressive training, in which the horse's load/speed are slowly increased, helps prevent bucked shins.
- Harder tracks and training surfaces also increase the likelihood of a horse getting bucked shins.
- The majority of Thoroughbred racehorses suffer from bucked shins during their first few months of training.

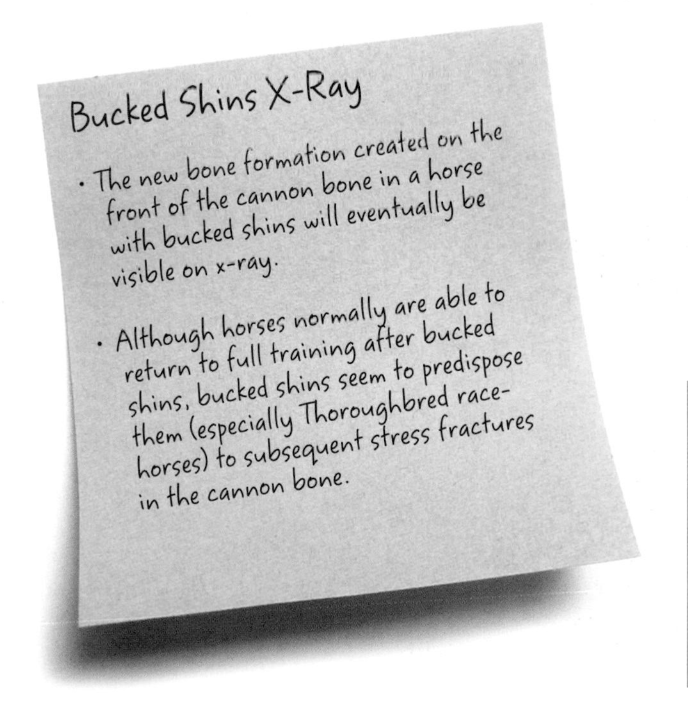

BURSITIS

Depending on the type, bursitis can be a minor blemish or a life-threatening emergency

To understand bursitis, it's important to first understand bursae. Bursae are small sacs that, like joint capsules, are lined with synovial membranes that produce synovial fluid. A true bursa sac is located between tendon and bone. When the bursa sac becomes inflamed, it's called bursitis.

To complicate matters, there are also acquired subcutaneous bursae that form where there wasn't a bursa sac before. These are also sometimes called a false bursa or a hygroma. Common conditions like capped elbows (shoe boils) and capped hocks are often cases of acquired bursitis. These types of chronic bursitis occur after repeated trauma causes the body to create a fibrous sac filled with fluid over the trau-

Shoe Boil

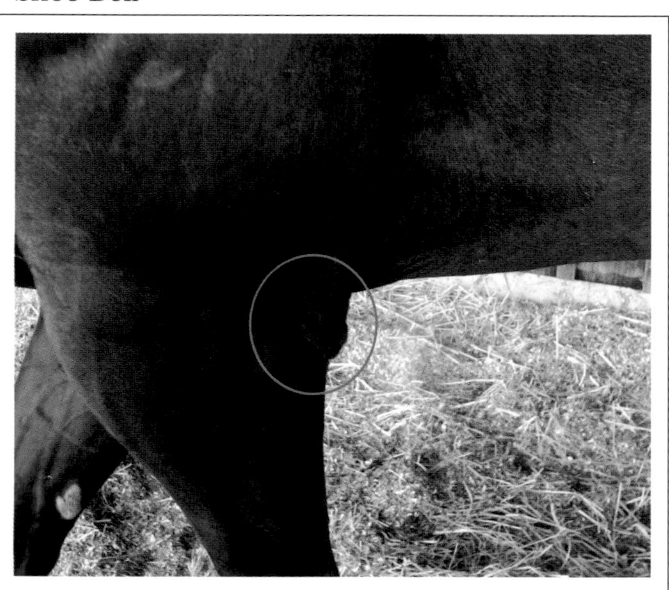

- *Capped elbow, olecranon bursitis,* or *shoe boil* usually refer to an acquired bursitis at the point of the elbow.
- This photo shows a healed shoe boil that was previously lanced, drained, and treated topically with DMSO and Fura-Zone.
- Although there is still a lasting cosmetic blemish, the shoe boil hasn't returned because the horse was outfitted with a shoe boil boot, and mats were added to the horse's stall along with plenty of shavings.

Shoe Boil Boots

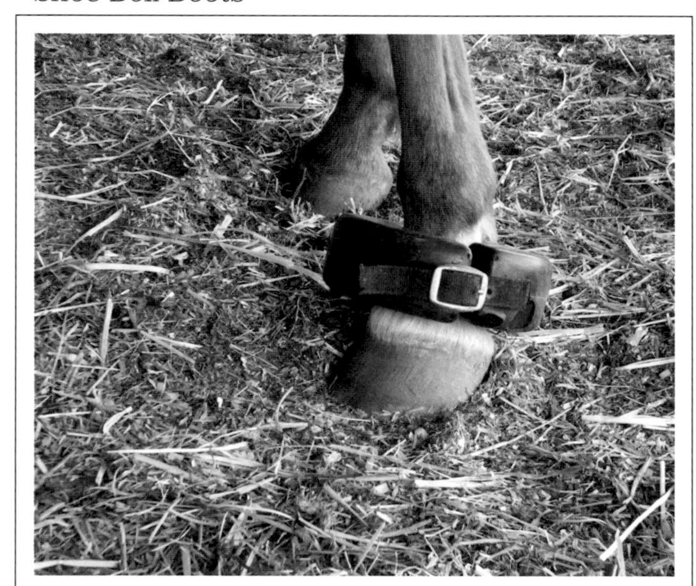

- The first signs of a shoe boil may be roughed up skin around the point of the elbow.
- To prevent shoe boils from reoccurring, worsening, or even appearing in the first place, purchase a large rubber shoe boil boot or "donut."
- Shoe boil boots, attached around the horse's pastern, prevent his shoe from coming into contact with his elbow when he lies down.
- The boot should be worn whenever the horse is unsupervised and checked regularly to be sure it's not creating chafing.

matized area. These are usually just blemishes

One example of true bursitis is bicipital bursitis, which is inflammation in the bursa sac at the top of the humerus below the biceps brachii tendon in the shoulder. True bursitis can be caused by a number of things, including direct trauma, tension from the bone and/or tendon caused by uneven loading to the area (poor conformation, shoeing) or overextension. Bone changes caused by osteoarthritis can also cause bursitis. These types of bursitis are usually accompanied by mild to moderate lameness, and although the swelling may not be easily visible, the affected area will likely be sore. Treatment varies by cause but will likely include decreased exercise, wrapping, icing, and anti-inflammatories injected and/or applied topically.

Septic bursitis caused by infection can result in severe lameness and is the most dangerous type of bursitis. One example, septic navicular bursitis, can occur when a nail penetrates the frog and navicular bursa causing infection. Prompt veterinary treatment including antibiotics is needed.

Capped Hocks

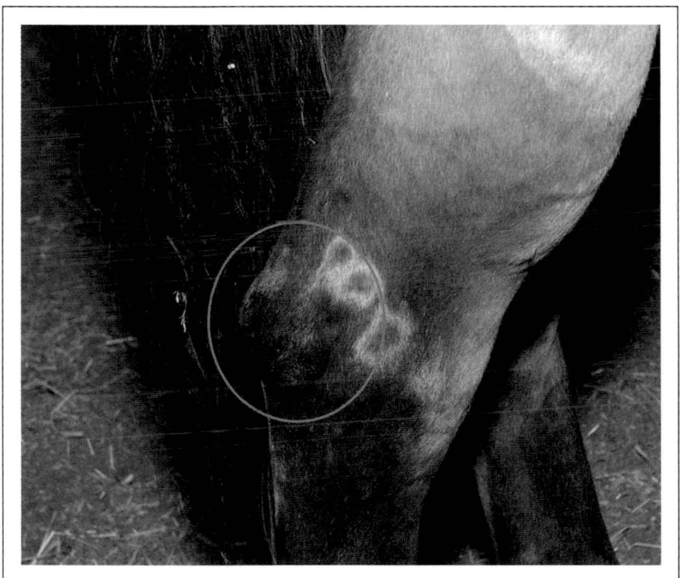

- Both capped hocks and carpal hygromas (acquired bursitis in the front of the knee) can be caused by kicks or the horse banging his legs against the trailer, stall walls, or fencing.
- Sometimes trauma can be treated with ice, rest, wraps, and anti-inflammatories before an acquired bursa fully forms.
- Capped hocks occur at the point of the hock.
- After chronic bursitis sets in, a capped hock is likely a long-term blemish that doesn't affect soundness.

Bedding

- A horse's knees and elbows can be traumatized by repeatedly coming into contact with hard-packed ground when the horse lies down.
- Both shoe boils/capped elbows and carpal hygromas (an acquired bursitis in the front of the knee) can be caused by the horse lying down on hard ground.
- To help prevent shoe boils and carpal hygromas or reduce the chances they will worsen or reoccur after treatment, provide your horse with a soft place to lie down or added bedding.

HEMATOMAS

Hematomas are blood filled swellings caused by direct trauma and are difficult to resolve

Horses are prone to injury and can sustain a knock to the leg when another horse kicks them, when they get tangled up in the fence, when they fall, or from a variety of other causes. Like hygromas/acquired bursitis, hematomas are fluid swellings. However, bursitis usually occurs over a bone, and hematomas often appear over a muscle mass, such as on the horse's thigh. If a horse's muscles or soft tissues are injured by direct trauma, they may begin to bleed, forming a hematoma under the skin.

Getting rid of the fluid swelling can be difficult, so if you catch a hematoma forming, apply ice and pressure to help the blood slow and clot. You'll also want to rest your horse

Knee Swelling

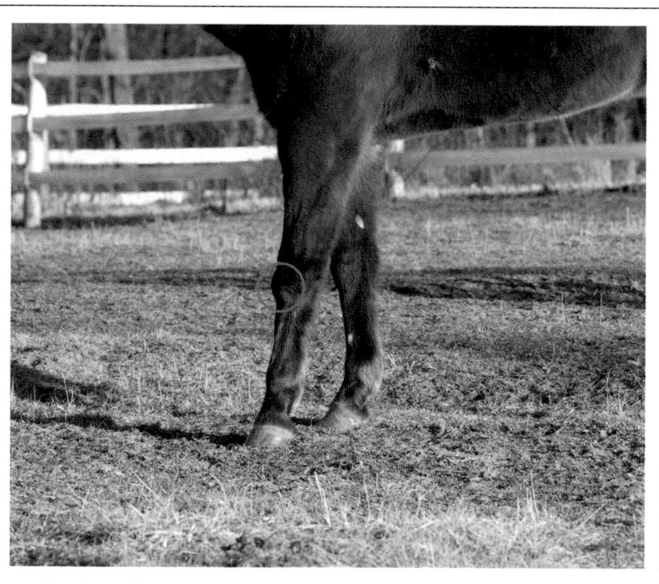

- Fluid swellings are often diagnosed as a hygroma (acquired bursitis), a seroma (serum-filled swelling), or a hematoma (blood-filled swelling).
- The vet may analyze the fluid for an accurate diagnosis.
- Draining the fluid is an option, but because there is usually a pocket left beneath the skin, it will often refill with serum.
- Steroid injections sometimes help, but it can be quite difficult to resolve fluid swellings.

Diagnostics and Treatment

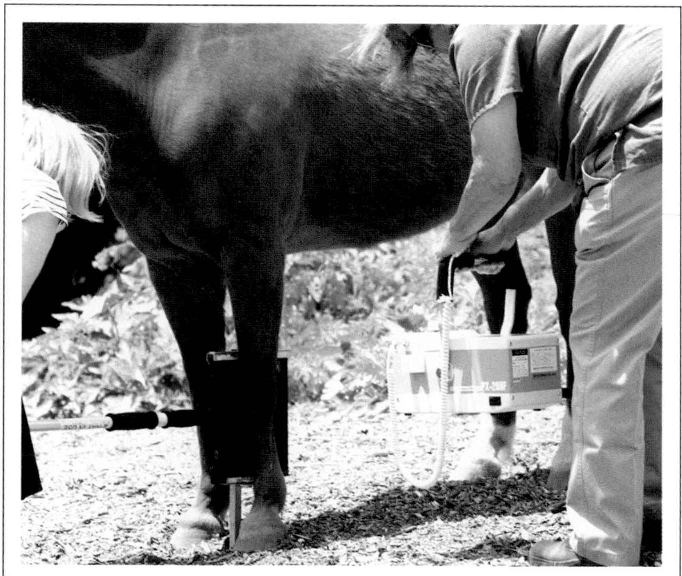

- Although a fluid-filled swelling such as a hematoma doesn't alone indicate x-rays, the horse's level of pain and the injury that caused the problem may cause the vet to recommend x-rays to examine whether any bones or joints are involved (healthy horse shown in photo).
- Old injuries and swellings are harder to treat than new ones, so it's always best to obtain an accurate diagnosis and treatment plan early.

during the early stages of hematoma formation and apply ice at regular intervals. Because other more serious conditions can look similar, have your vet examine the horse to determine if you are indeed dealing with a hematoma. Depending on the location, the vet may suggest ultrasound or x-rays to see how deep the hematoma is and whether bones, muscles, or joints are involved.

The vet will not want to drain or open a hematoma during the first few days, as the injury will likely still be bleeding. Most of the fluid will eventually be reabsorbed by the body, but some fluid swelling may remain, which is why owners and vets sometimes opt to eventually drain the hematoma. However, draining carries a risk for infection, and the area may simply refill with serum. Repeated draining increases the likelihood of infection. If the hematoma is not interfering with the horse's movement or his equipment, it may be best to leave it alone.

Swellings: When to Call the Vet

- There's a fine line between calling the vet over every little thing and not getting the vet involved early enough.
- If a leg swelling is mild and not accompanied by heat or pain (on movement or palpation), then many owners will treat it with a standing or pressure wrap, rest, and cold therapy.
- However, swelling that doesn't improve quickly with conservative treatment, or swelling that is accompanied by heat or pain should always prompt a call to the vet.

Trauma

- When horses bicker, bites and kicks often cause injury.
- Hematomas often form after a kick from another horse.
- Reduce the chances of fighting by introducing horses slowly, providing plenty of space, and feeding them separately or far apart.
- While a certain amount of bickering can be expected to establish dominance, each horse's personality should also be assessed before deciding if a pair will make good pasture buddies.

WOUND CARE AND HEALING

Caring for wounds is part of horse ownership, but leg wounds need special consideration

Horse owners often joke that if there's a way for a horse to hurt himself, he will, and legs are especially vulnerable. It's best to consult the vet when dealing with all but the most minor leg wounds, as wounds to the legs can easily affect joints, tendons, ligaments, and nerves even if the horse appears sound. Wounds along the sensitive coronet or heel bulbs also need veterinary attention and can affect future hoof growth.

Wounds are categorized into different types: Abrasions are when the horse has scraped off his hair and the top layers of skin. Puncture wounds are often small and deep, involving sharp objects. Because of the risk for infection, puncture

Cleaning Wounds

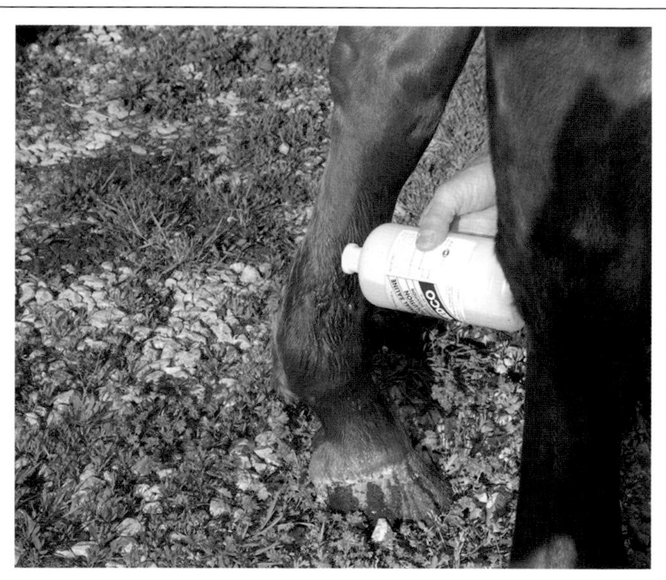

- A sterile saline solution is best for cleaning wounds, as it's gentle and effective.
- You can also make your own saline solution by mixing a teaspoon of salt per quart of water.
- Put the saline solution in a large syringe, and gently squirt the wound until dirt and debris have been washed away.
- Opinions vary on using topical wound treatments (antiseptic and antibiotic ointments and sprays), so ask your vet what you should stock in your first-aid kit.

Wrapping

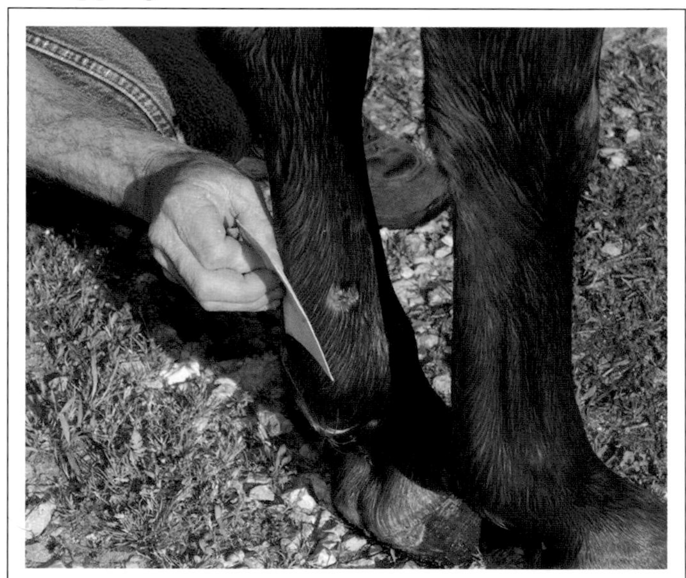

- Wrapping for wounds is similar to applying standing wraps (detailed on page 188) but with slightly different materials.
- First, apply a nonstick pad over the wound. The pad will usually stay in place if any topical products have been applied to the wound.
- Instead of washable cotton padding, disposable cotton batting is often used for padding around wounds.
- Apply self-adhesive wrap over the batting. Duct tape can secure the end, but never wrap duct tape all the way around a leg, which can impede circulation.

wounds need veterinary attention. Lacerations are cuts that go through all the layers of skin. Lacerations usually need a vet consult and can often benefit from sutures.

Bleeding wounds may require immediate bandaging. While you wait for the vet, apply pressure over the bandaged wound using your hand or fingers. You can also apply ice over the bandaged wound or elevate the leg if the horse will allow it and stand quietly.

When you encounter a fresh wound that's not actively bleeding, flush the wound with a saline solution. This is the best way to gently clean the area. If the vet is coming, wait for further instructions. If it's a minor wound you'll treat yourself, you can apply any topical treatments your vet recommends. For wounds in areas that can be wrapped, follow the instructions on page 106.

It's important to have a well stocked first-aid kit ready at all times. The kit will include items for treating wounds and bandaging legs. See page 210 for a detailed list of items to include in your first-aid kit.

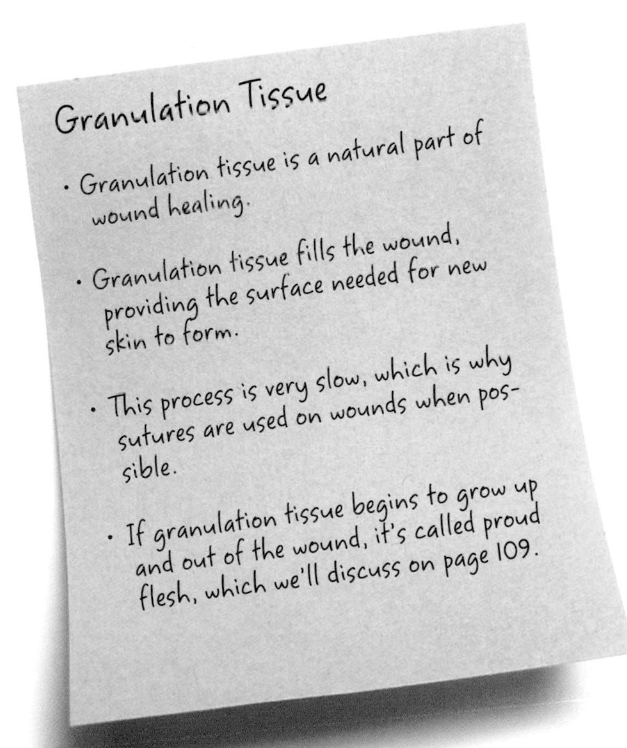

Sutures

- Some lacerations can be sutured, making healing quicker and easier, and possibly reducing scarring.
- If the sides of a wound can't be brought together, then the wound is not a candidate for stitches.
- The vet may suggest restricted movement and offer wrapping instructions to help sutures hold over high-motion areas like joints.
- The vet will either schedule a time to remove the stitches or use dissolvable stitches.

WOUND COMPLICATIONS

Proud flesh and infection are two common complications to wound healing

Wounds must be kept as clean as possible during healing, which is why, when possible, wrapping is the best bet. Brief and gentle daily hosing or flushing with saline solution may be helpful to keep unwrapped, minor wounds clean. Keep flies away by applying repellent around but not on the wound.

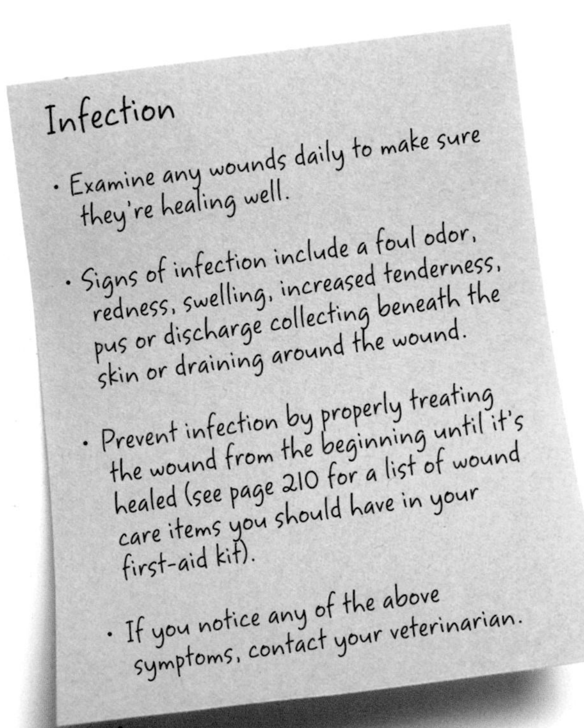

Wounds that are not properly cleaned or cared for during healing, or that had a great deal of dirt and debris in them to begin with, may become infected. Infections always require prompt veterinary treatment. Observe your horse and inspect his wound daily.

Signs of infection include the wound draining pus or cloudy

Fever

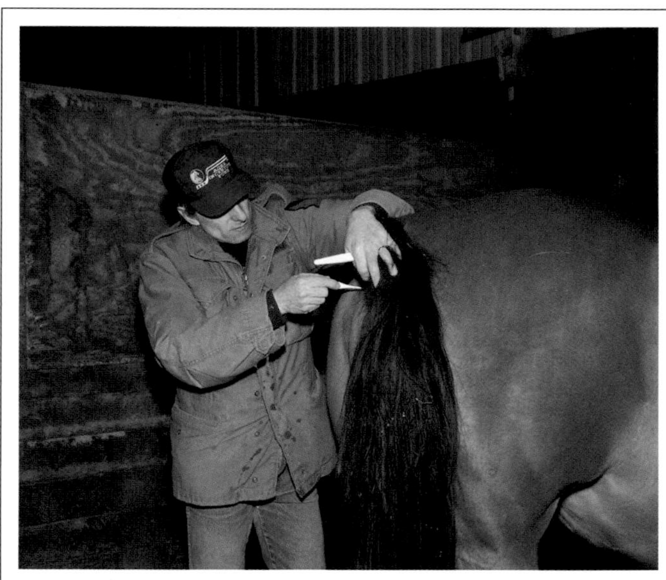

- A horse's normal temperature is 99.5 to 101.5 degrees Fahrenheit, and an elevated temperature is one sign of infection.
- To take your horse's temperature, use a lubricant like petroleum jelly on the tip of a digital thermometer before inserting the tip rectally.
- To avoid losing the thermometer in or out of the horse, attach a string and a clip for clipping to his tail.
- Digital thermometers usually only take a minute and make a small beep when they're done.

fluid, increased redness or irritation around the wound, increased pain or swelling (even after the wound is two days old or older), or fever. If severe, infection may also cause the horse to stop eating or drinking or be reluctant to move. The veterinarian will most likely prescribe antibiotics to fight the infection.

Another less serious complication, but one that still requires veterinary attention, is proud flesh. Granulation tissue naturally fills in the wound and has tiny blood vessels that help feed the new skin growth. However, when granulation tissue runs amok and grows in excess, it's called proud flesh and halts natural wound healing. Limbs are especially prone to proud flesh because their constant moving makes it hard for the skin to come back together over the wound, and the granulation tissue may be pushed out.

There are a number of over-the-counter products to combat proud flesh aimed at drying the tissue up or eating away at it, which ultimately does more harm to the healing process. The best option is to have your veterinarian trim the excess granulation tissue and prescribe a topical corticosteroid antibiotic.

Proud Flesh

- Any time the granulation tissue grows up and out of the wound rather than just filling in the wound, you have a problem with proud flesh and should consult your veterinarian.
- Proud flesh is excessive granulation tissue that looks a bit like pink cauliflower.
- Rather than helping new skin form as granulation tissue is supposed to do, proud flesh prohibits the wound from properly closing and healing.

Scarring

- Bigger wounds take longer to heal and are more likely to leave larger scars.
- Scars are often minimized if the wound can be sutured.
- Wounds that heal without complication leave smaller scars than wounds that had trouble healing.
- Scars are often hairless, or the hair grows in white. This is because the wound permanently damaged hair follicle cells, or in the case of white hair, the tissue's pigmenting capabilities were lost.

CELLULITIS AND LYMPHANGITIS

Cellulitis and lymphangitis cause swelling to one or more legs and require veterinary care

Cellulitis and lymphangitis are two types of inflammation that can occur after even a minor wound to a horse's leg. Cellulitis is the more common of the two and is an infection of the soft tissues. It usually occurs in just one leg—the leg with the wound—and is most often due to bacteria entering through the wound. Cellulitis may start as swelling around the wound, and the swelling then spreads up or down the leg. Any time a leg is swollen, it's best to consult your vet. Antibiotics are often recommended, along with proper wound treatment, cold therapy, and wraps.

Lymphangitis is inflammation of the lymphatics (lymph vessels) due to a disruption in the lymphatic flow. Like cellu-

Cellulitis

- Cellulitis usually affects only one leg, and the swelling can move up or down the leg, even affecting the area above the horse's knee or hock.
- There are a number of conditions that can cause swelling in the legs, so consult your veterinarian.
- Cellulitis can affect the bones or tendon sheaths.
- In addition to bacteria entering through an external wound, a joint, bone, or tendon sheath abscess may also cause cellulitis.

Lymphangitis

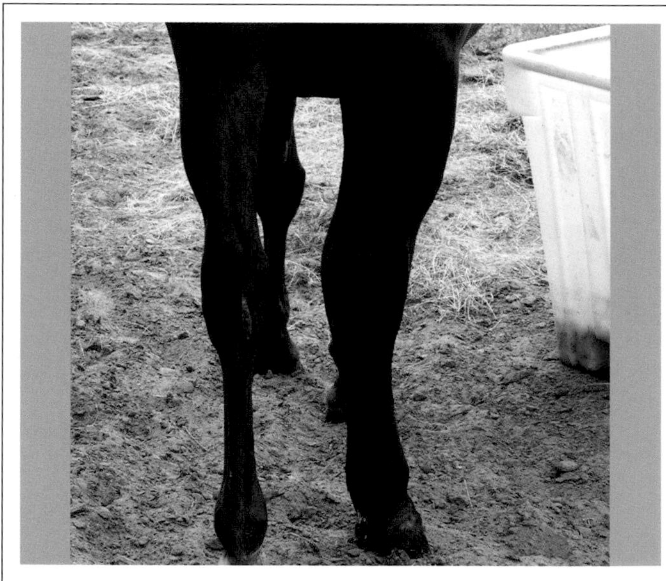

- Lymphangitis can cause swelling from the stifle down, the hock down, or the fetlock down. In the above photo, you can see this mare's entire right hind leg is severely swollen.
- Swelling in the lower leg can become so severe the leg doubles its normal size.
- Because of the swelling, it is often difficult for the horse to move normally or take more than a few steps.
- Lymphangitis is sometimes accompanied by fever.

litis, lymphangitis can be caused by a bacterial infection that enters through a wound.

Although there is little known about it, there is also an endemic type of lymphangitis that can affect a number of horses in the same stable. This is more likely to occur in the wet season; maintaining sanitation helps prevent the problem. It is usually difficult to determine the exact cause of lymphangitis.

Lymphangitis is more likely to affect the hind legs than the front legs and can result in swelling from the lower leg down, the midleg down, or the entire leg. When bacteria cause lymphangitis, serum can leak out of the swelling or form draining ulcers. This is called ulcerative lymphangitis.

The vet will likely prescribe appropriate antibiotics and possibly anti-inflammatories, along with hydrotherapy and rest. However, after a horse has had a severe bout of lymphangitis, permanent fibrous scarring tissue may remain or his lymphatic flow may be altered, resulting in reoccurrence.

Lymphangitis Case History

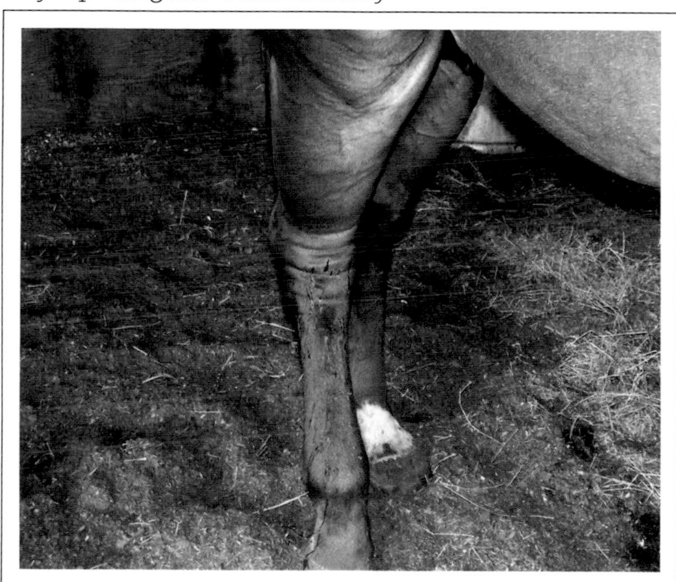

- This photo shows a horse with ulcerative lymphangitis.
- Before swelling was noticeable, the mare showed signs of discomfort similar to colic, but just a few hours later, her hind leg was severely swollen, and she was reluctant to walk.
- The mare was promptly placed on antibiotics, bute, and hydrotherapy.
- The following day the weeping through the skin was noticeable along with increased swelling. Both symptoms were beginning to subside by the next day.

Housing

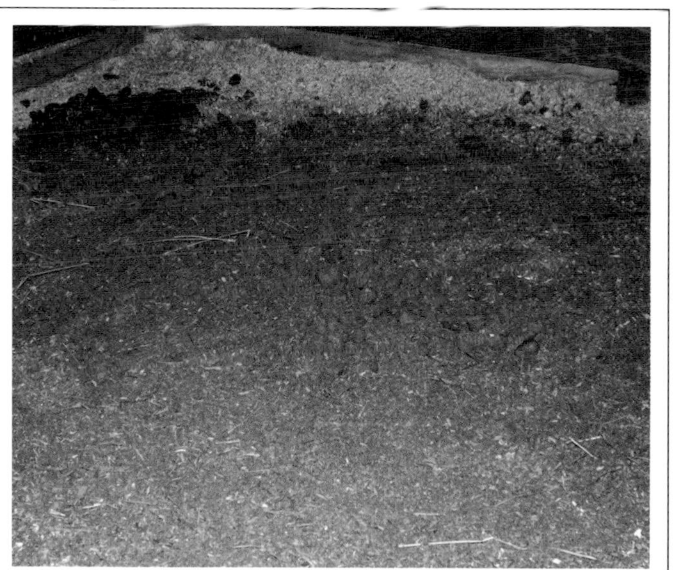

- Your horse spends most of his time in his stall or pen, so his housing has a major impact on his health.
- Unsanitary conditions, such as stalls that aren't kept clean and dry, provide a prime environment for bacteria, which can cause cellulitis and lymphangitis.
- Stalls should be cleaned twice daily.
- Even if you can't groom your horse daily, visually inspect his legs when you clean his pen so that any problems can be caught early.

HOOF PUNCTURE WOUNDS

Hoof puncture wounds are somewhat common and should always be treated seriously

Hoof puncture wounds occur when a horse steps on a nail, staple, stiff wire, or other sharp object, and it penetrates his sole or frog. How deep the object goes, what internal structures it damages, whether infection develops, and whether it's treated early all affect the outcome. Shallow puncture wounds may not cause much of a problem, but deep puncture wounds or infections can cause severe lameness leading to euthanasia.

Beneath the sole and frog are the coffin joint and bone, navicular bursa sac and bone, and the digital flexor tendon and tendon sheath. Damage or infection to any of these areas is extremely serious. Therefore, it's important to recognize

Puncture Wound

- It can be hard to find the hole once an object is removed since the sole and frog are somewhat soft.
- If you discover a penetrating object and must remove it before the vet arrives to avoid it sinking deeper, note how deep the object was in, what direction it went, and where it was located.
- Bacteria can be introduced by the penetrating object or through the wound, so prompt veterinary care is always necessary.

X-Ray

- If you can leave the object in without risk of it traveling deeper, the vet can x-ray the hoof to see how deep the penetration is and what structures may be involved. Otherwise a contrast solution may be injected into the site to visualize the tract.
- Whether the object is in the hoof or removed, if it was likely a deep puncture, the vet may recommend additional diagnostics such as MRI to determine what parts of the hoof were damaged.

puncture wounds early and treat them promptly.

If the object your horse stepped on is still in his hoof upon your inspection, then it will be easier for the vet to determine where the wound occurred, how deep it is, and what direction the object traveled. However, many times leaving the object in risks driving it deeper. If you must remove it, use a hoof knife to pare away a bit at the entry site, or circle the site with a permanent marker. If you don't mark the area, it may close up or be impossible to locate.

If the object is no longer in the hoof, you may not know your horse suffered a puncture wound. In this case, the first sign of trouble is mild to moderate lameness stemming from the lower leg. Call the vet to help you determine the cause.

The vet's goal when treating a puncture wound will be to determine if there's infection and what, if any, structures were penetrated. Next, the vet must thoroughly clean the area and create a draining tract. If the puncture wound was deep or compromised vital internal structures, surgery to clean and treat the area may be necessary.

Healing

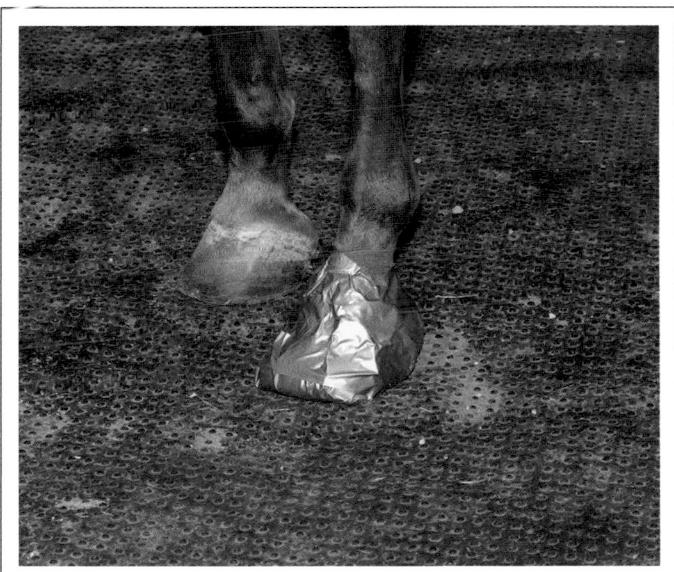

- After the puncture wound area has been cleaned and a draining tract created, or after surgery, protect and keep the hoof clean while it heals.
- The vet may suggest the horse wear a bandage, a rubber hoof boot, or a hospital plate shoe until the hoof heals.
- Bandages, rubber boots, and hospital plate shoes protect the wound and allow access to treat it and check on its progress.

Symptoms

- If the penetrating object is not found in the hoof, the first sign of a possible puncture wound will likely be lameness.
- Later signs can include heat in the hoof and lower leg, which can be felt by comparing one hoof to another.
- Swelling and a strong digital pulse may also occur in the lower leg after a puncture wound.
- An abscess can eventually form in an untreated puncture wound and may erupt.

TENDON AND LIGAMENT INJURY

It's important to recognize tendon and ligament problems, as prompt treatment creates a better outcome

Tendons and ligaments are made of fibrous bundles. Ligaments join bone to bone, keeping the bones from overflexing or overextending. Tendons join muscle to bone, and when the muscle contracts, the tendon transfers that contraction to bone. There are extensor tendons that bring the legs forward and flexor tendons that flex it back.

Tendons and ligaments can stretch or tear through repeated damage or a single episode of trauma or overload. Often when tendons and ligaments are injured, their fibers are torn, creating lesions. Any tendon or ligament can be injured.

Injuries are graded type one through type four. The more fibers torn and the bigger the lesion, the worse the injury. Type

Trauma

- Tendons and ligaments play a vital role in movement and weight bearing, and damage can occur over time as use trauma or in one traumatic episode.
- Think of tendons and ligaments as ropes, with a great deal of weight to bear—parts of the rope can break suddenly, or fraying can happen gradually.
- You can also compare tendons and ligaments to rubber bands—they can only take so much overuse before they become stretched out or break.

Inflammation

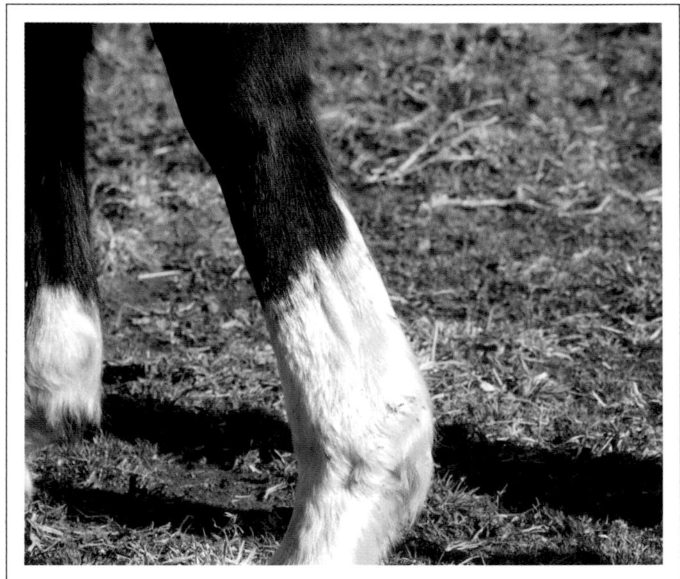

- Once tendon or ligament fibers are disrupted by injury, inflammation causes further disruption, which is why early detection and treatment sre key to a quicker recovery.
- The amount of inflammation or lameness doesn't always correlate with the severity of the injury, so consult your veterinarian for an accurate diagnosis and appropriate treatment plan.
- While the fibrous tendon and ligament tissues can heal after injury, they will not be as strong as the original tissues.

one injuries are the least severe. Type four injuries affect most of the cross-sectional area and are considered a rupture.

The term *bowed tendon* comes about from the bowing out that occurs at the back of the leg with certain common tendon and ligament injuries. Laymen classify bows as high, middle, or low, referring to their location on the leg.

If you notice any swelling, heat, or lameness in your horse's leg, call the vet. To diagnose a tendon or ligament injury, the vet will first palpate the area and then confirm the diagnosis with an ultrasound or MRI to visualize the soft tissues.

Horses with poor conformation, such as long, sloping pasterns or long toe/low heel place added strain on tendons and ligaments. Horses not properly warmed up, not properly conditioned, or worked on deep or slippery footing are also more prone to tendon and ligament injuries.

Common treatments are discussed below. Some chronic injuries benefit from desmotomy (incision to the ligament) or tendon splitting (see page 126).

Ultrasound

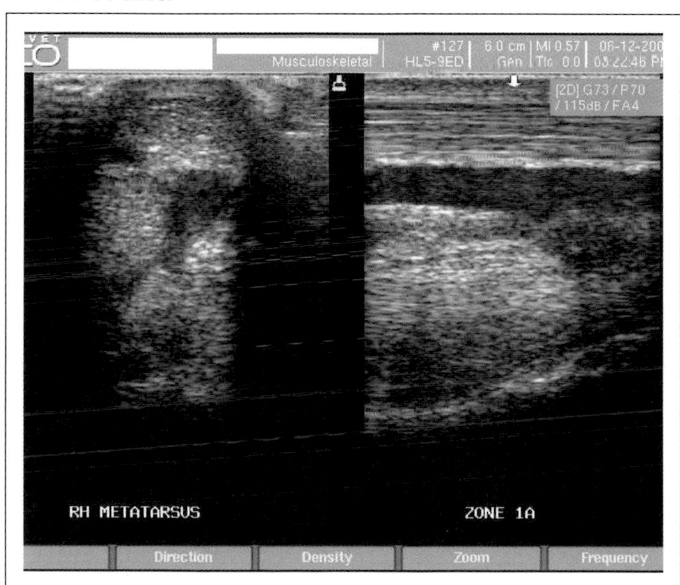

- Ultrasound is the most common tool in diagnosing tendon and ligament problems and visualizing where the problem is and how severe the damage.
- Ultrasound is often used throughout healing to monitor progress.
- Before returning the horse to work, ultrasound can confirm adequate healing.
- This ultrasound is of a proximal suspensory ligament core lesion on a nineteen-year-old jumper. This horse retired from jumping.

Treatment

- Wraps applied correctly and with plenty of padding are often needed for support and to reduce swelling. The vet may also recommend DMSO application.
- Icing is helpful for thirty minutes two to four times a day during the first seven to ten days.
- Oral anti-inflammatory drugs and hyaluronate or polysulfated glycoaminoglycan injections may be prescribed by the vet.
- Return to exercise will be very gradual: stall rest, then hand walking, then walking under saddle, and so on.

FLEXOR TENDON INJURY

The superficial digital flexor tendon is the most commonly injured tendon, followed by the deep digital flexor tendon

The superficial digital flexor tendon (SDFT) and the deep digital flexor tendon (DDFT) are responsible for flexing the horse's leg during movement and helping to stabilize the leg during weight bearing. Both travel down the back of the horse's leg, with the SDFT closer to the surface. Tendinitis is strain-induced inflammation in the flexor tendons.

The SDFT is more commonly injured than the DDFT. This may be in part because the SDFT bears weight in the early phase of the gallop before the DDFT shares the load. During the gallop, the fetlock lowers toward the ground and may even touch the ground, which is why SDFT injuries are common in horses. It's common for damage and microtears to

SDFT Injury

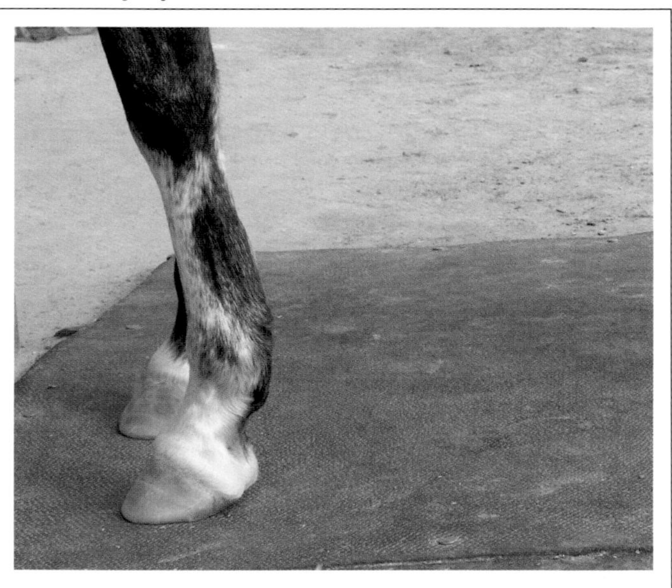

- Mild swelling and heat may be the first sign of tendon injury, even before lameness is apparent. Early detection and treatment are key to a positive outcome.
- The front legs are more likely to suffer from tendon injuries, since they bear the most weight.
- SDFT injuries most often occur in the mid–cannon bone area, where the SDFT has the smallest cross-sectional area.
- This area may also be prone to degeneration, predisposing it to injury.

SDFT Injury Recovery

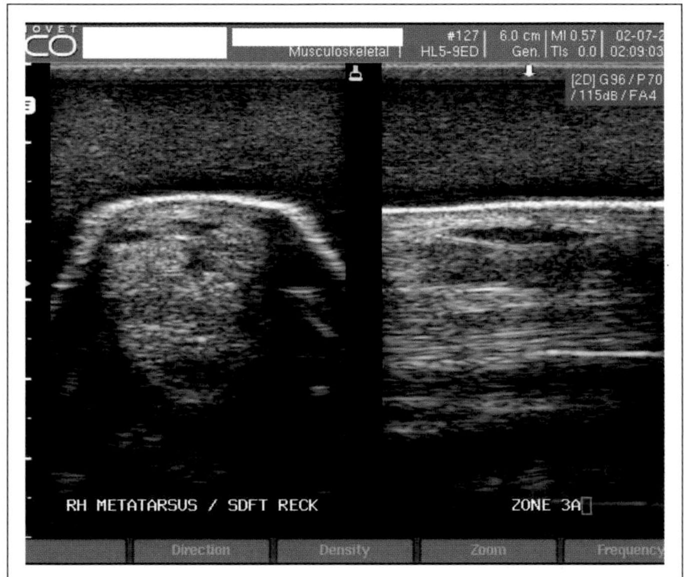

- This ultrasound shows a SDFT lesion.
- The mare's hind limb SDFT was lacerated by the hoof of another horse. She was brought to the hospital for surgical repair of the laceration.
- Although the mare has a permanent defect of the SDFT, she is sound for trail use.
- Stall rest is recommended after a tendon injury, followed by hand walking and then walking under saddle.

happen over time, leading up to the injury.

Injuries to the SDFT can be mild or severe with complete rupture. Although less common, the DDFT can also be injured with varying degrees of severity. The DDFT ends lower than the SDFT, connecting to the coffin bone, and injuries to the DDFT are most common at the fetlock level or below.

Immediately following injury, there is inflammation and swelling caused by hemorrhage and the formation of an intratendinous hematoma. Then scar tissue formation begins. Treatment goals are to reduce inflammation and scar tissue and to restore function.

During the acute stage following injury, icing is recommended for thirty minutes two to four times a day. A pressure bandage should be applied between icings, and the vet may recommend DMSO application. Bute (phenylbutazone) will likely be prescribed orally, and the vet may also give hyaluronate or polysulfated glycoaminoglycan injections. Severe injury to either flexor tendon can result in the fetlock dropping farther toward the ground than usual, and may require a plaster or gel cast worn during healing.

DDFT Injury

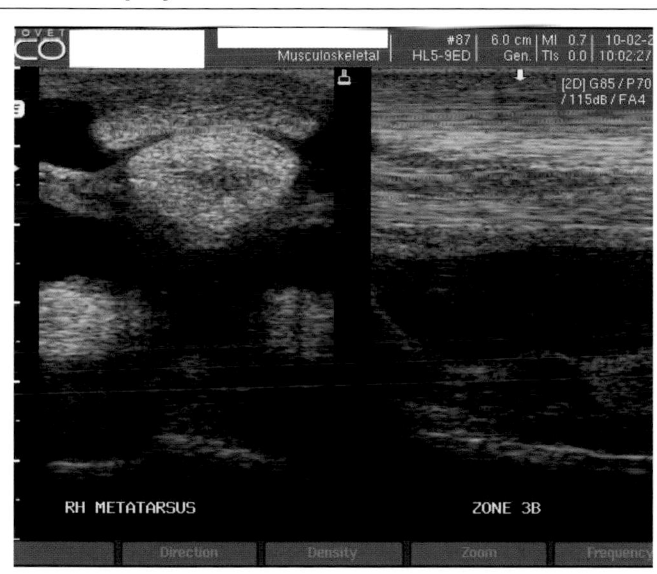

- This ultrasound shows a DDFT lesion. The acute injury occurred during a trail ride.
- This particular injury healed with calcification of the ligament, and the horse is now sound and in low-level exercise.
- Some severe DDFT injuries can end a horse's career and cause persistent lameness.
- Owners can help prevent re-injury to a tendon by maintaining balanced hooves, and by monitoring footing and workload accordingly.

Extensor Tendon Injury

- The extensor tendon runs along the front of the leg and connects to the top of the coffin bone.
- The extensor tendon's job is to extend the leg forward during movement.
- Injuries to the extensor tendon are less common than flexor tendon injuries.
- When injured, extensor tendons tend to heal more quickly and have a better prognosis than flexor tendons.

SUSPENSORY LIGAMENTS

Injury to the suspensory ligament is likely the most common soft-tissue injury

The suspensory ligament starts at the upper back part of the cannon bone and travels down the back of the leg splitting into two branches (medial and lateral) that insert into the proximal sesamoid bones at the fetlock. Two small branches of the suspensory ligament then travel toward the front of the leg—one on each side. Smaller ligaments also travel down the back of the short and long pastern bones and are called the distal sesamoidean ligaments.

The suspensory ligament is an important part of the suspensory apparatus that tries to keep the fetlock from overextending, but injuries can occur when the fetlock hyperextends. Suspensory ligament injuries are the most common

Proximal Suspensory Desmitis

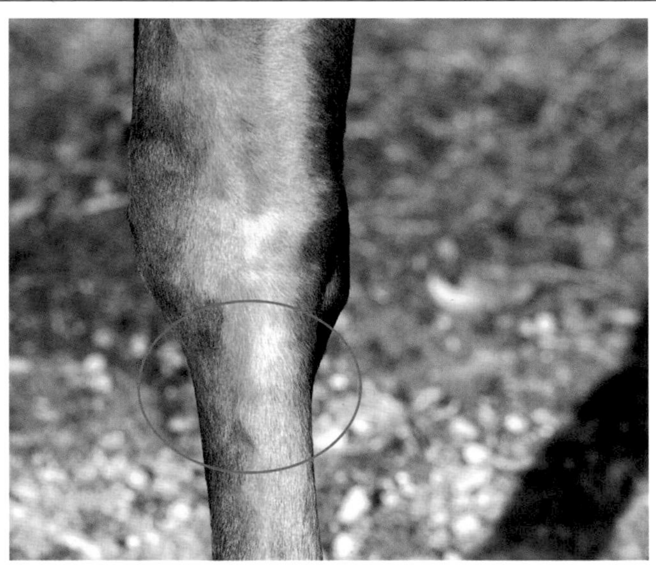

- Proximal suspensory desmitis (injury to the top portion of the suspensory ligament as indicated by the circle) can be hard to detect because the ligament is deep within the tissues there.
- The vet may use nerve blocks, ultrasound, and nuclear scintigraphy to pinpoint the injury.
- Proximal suspensory desmitis often shows more subtle lameness and swelling than lower injuries, but healing time is usually lengthy.
- Severe suspensory ligament injuries make re-injury more likely.

Suspensory Body Desmitis

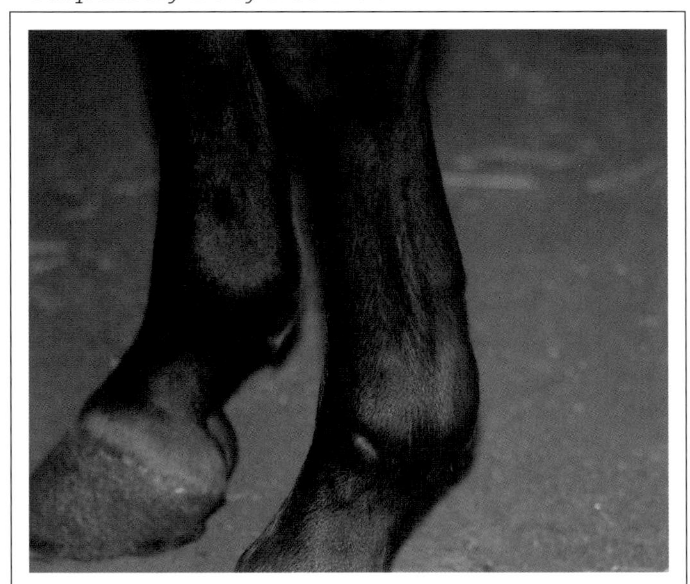

- If suspensory body desmitis becomes chronic, the ligament will thicken, and the horse may be haunted by lameness throughout his life.
- In horses suffering from more severe suspensory ligament injuries than the one pictured in the photograph above, the fetlock can drop down farther than it should.
- Suspensory ligament injuries to the front legs generally have a better outcome than hind leg injuries.

tendon or ligament injury. Desmitis refers to inflammation or damage to a ligament, and injuries to the suspensory ligament are categorized based on where they occur.

Proximal suspensory desmitis refers to injury near the origin of the ligament, at the top, back part of the cannon bone. Suspensory body desmitis means injury to the middle portion of the ligament, before it splits into medial and lateral branches. This is the most common place for injury to occur.

The medial and lateral branches of the ligament can also be vulnerable because they have a smaller cross-sectional area. Suspensory branch and distal sesamoidean ligament injuries can be accompanied by injury to the proximal sesamoid bones. Sesamoiditis is pain associated with the proximal sesamoid bones, and signs include lameness, heat, and swelling, or thickening around the fetlock.

Treatment for suspensory ligament injuries is similar to that for tendon injuries. Healing will likely take several months. Ultrasound repeated every thirty to sixty days can monitor healing and check for re-injury. Prognosis depends on treatment, and on the severity and location of the original injury.

Suspensory Branch Injuries

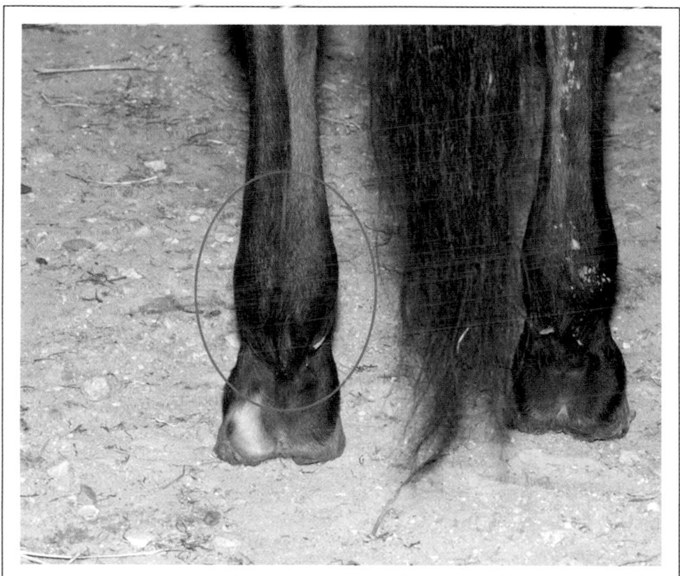

- At the proximal sesamoid bones, the suspensory ligament has two branches that travel forward medially and laterally (one on each side) to join the common digital extensor tendon in front.
- Injury to the medial or lateral branches of the suspensory ligament usually produces visible swelling, and the vet is able to easily palpate the area.
- Poor foot balance can lead to suspensory ligament branch injury.

Sesamoidean Ligament Injuries

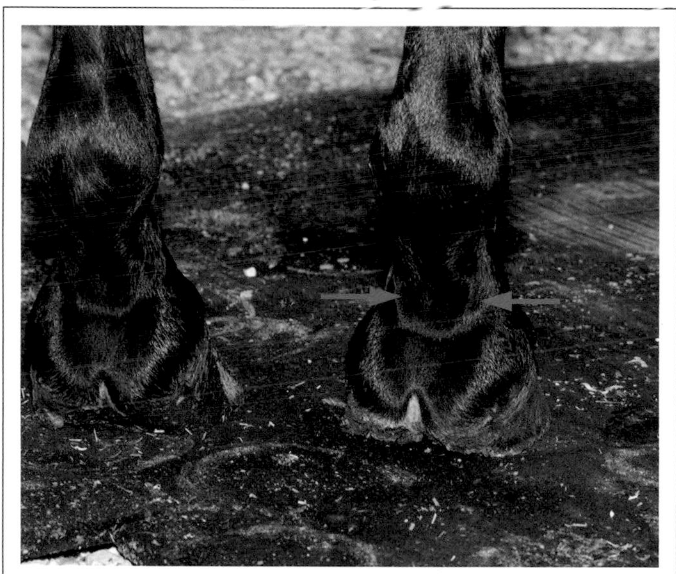

- Distal sesamoidean ligament injuries can go unnoticed, and diagnosis can be difficult because they're deep within the tissues (arrows show location in the photograph above).
- Treatment often includes rest, ice, wraps or poultices, and bute (phenylbutazone).
- Sesamoidean ligament problems can become chronic, but prognosis depends on severity, location, and quick detection and treatment. Horses may need six months before returning to work.
- Corrective shoeing to ease breakover can be helpful.

TWO COMMON LIGAMENT INJURIES

Inferior check ligament injuries and upward fixation of the patella are unrelated but can both cause lameness

There are two relatively common but unrelated ligament issues. The first are injuries to the inferior check ligament. The inferior check ligament starts at the back of the knee and joins the deep digital flexor tendon midpoint along the cannon bone. Older horses are most at risk for inferior check ligament injuries, possibly due to how the ligament materials and fibers change with age. These injuries are also somewhat common in jumpers. Inferior check ligament injuries may cause more swelling than lameness, and as always, early treatment is key.

Ultrasound will confirm an inferior check ligament injury, and treatment is similar to other tendon and ligament issues:

Check Ligaments

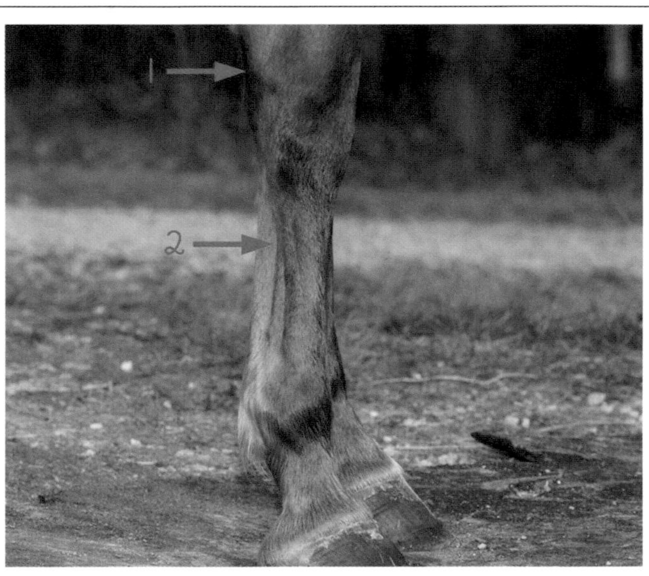

- There is more than one check ligament, as shown in the photograph above.
- The superior check ligament (accessory ligament of the superficial digital flexor tendon) connects to the superficial digital flexor tendon (arrow 1).
- The inferior check ligament (accessory ligament of the deep digital flexor tendon) connects to the deep digital flexor tendon (arrow 2).
- The inferior check ligament helps stabilize the knee (carpus) and shares part of the deep digital flexor tendon's load.

Inferior Check Ligament Injuries

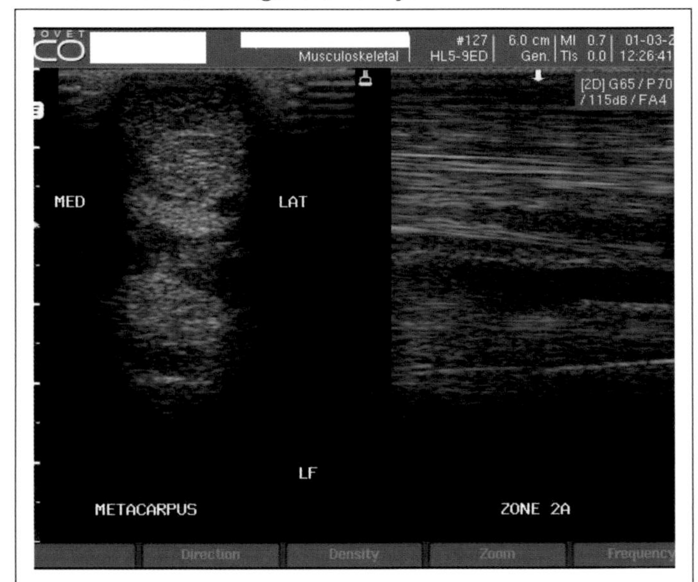

- This ultrasound shows an inferior (also called distal) check ligament lesion.
- The swelling associated with an inferior check ligament injury will be closer to the cannon bone than an injury to the superficial flexor tendon, which often shows up as a classic bow.
- While the injury heals, carefully controlled, gradually increasing exercise is recommended (not turnout).
- Not all horses can return to their previous athletic careers after an inferior check ligament injury, and some thickening in the area may remain.

icing during acute inflammation phase with pressure wraps in between, anti-inflammatories, and possibly hyaluronate or polysulfated glycoaminoglycan injections. Healing may take six months or more, and ultrasound will confirm when a horse is ready to resume work.

The second issue, upward fixation of the patella, is a hind-end only ligament issue. The horse's stifle joint sits below his femur (thigh bone) and is comparable to the human knee. The patella is like a kneecap, and it slides along the femur during flexion. Horses have three distal patellar ligaments, and sometimes the medial one gets hooked on the femur. This is most common in horses just beginning training that have poor muscle tone and/or may have a more lax ligament due to their poor conditioning.

This condition can cause either or both legs to lock during extension, with the toe pointing down. It may lock for less than a second or in severe cases, the horse may require assistance to unlock it. In less severe cases, a mild hind-end lameness may be the only sign, with the horse demonstrating a shortened stride or dragging his toes.

Upward Fixating Patella Diagnosis

- Horses demonstrating the classic locking during extension of one or both hind legs are relatively easy to diagnose with upward fixation of the patella, but horses with more subtle symptoms may require X-ray and ultrasound for a definitive diagnosis.
- The problem may only occur at the beginning of a workout, and popping sounds are sometimes heard.
- Horses with straight hocks, cow hocks, or base narrow conformation are more likely to suffer from upward fixation of the patella.

Upward Fixating Patella Treatment

- Affected horses may not want to canter, canter on the wrong lead (shown in the photograph above in a healthy horse) or swap leads.
- Improving muscle tone with exercise is often the first treatment, along with corrective shoeing.
- In stubborn cases, the vet may prescribe estrogen to geldings to change the tension of the ligaments or suggest blistering (irritating) the ligament to shorten it by creating an inflammatory response. Cutting the medial ligament is sometimes required.

ADDITIONAL LIGAMENT ISSUES

The fetlock's palmar-plantar annular ligament can cause trouble, as can ligaments in the hock and knee

Injury can affect any tendon or ligament in a horse, but this chapter focuses on the most common tendon and ligament injuries. Thus far, we haven't looked at any annular ligament injuries. Annular ligaments are wraparound ligaments. One of their jobs is to help tendons stay in place and function properly. Trouble can occur in the palmar-plantar annular ligament (PAL) in the high-motion fetlock joint. Palmar generally means lower back part of the front leg (or the back side of a leg), and plantar means lower back part of the hind leg. There are two annular ligaments lower than the PAL, the proximal digital annular ligament just below it, and below that the distal digital annular ligament.

Palmar-Plantar Annular Ligament

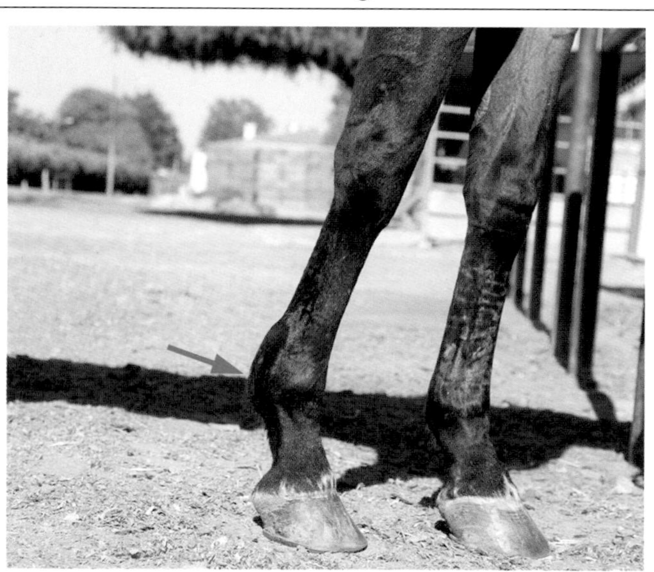

- Thickening of the palmar-plantar annular ligament or tendonitis of the DDFT or SDFT can cause constriction in the area. The horse was diagnosed with a constricted annular ligament and underwent surgery to correct the problem.
- Sometimes both palmar-plantar annular ligament thickening and tendonitis are present.
- Constriction in this ligament area causes pain and lameness. Constriction can also lead to tenosynovitis and distention of the digital flexor sheath, which will be visible.

PAL Diagnosis

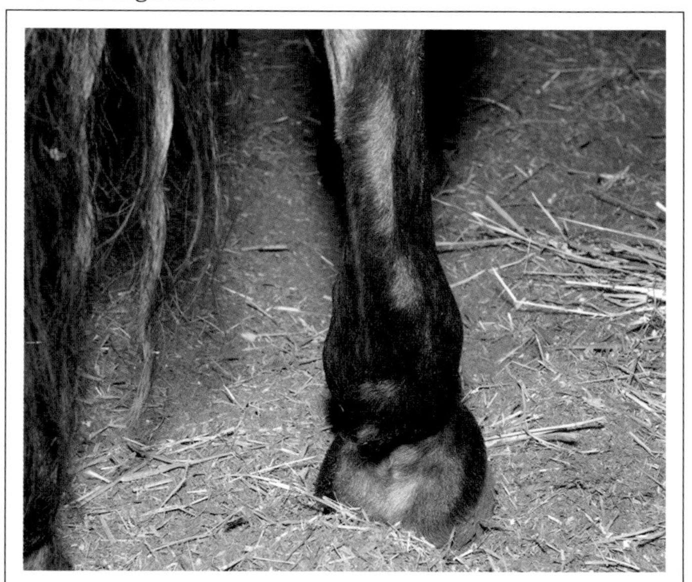

- Palmar-plantar annular ligament problems can be difficult to diagnose.
- The palmar-plantar annular ligament is not normally easy to visualize via ultrasound.
- However, ultrasound does help the vet visualize other soft tissues and a thickened palmar-plantar annular ligament.
- Diagnosis of palmar-plantar annular ligament constriction may also include the use of nerve blocks and x-rays.

The PAL goes across the back of the fetlock joint between the proximal sesamoid bones and helps to form a canal that the deep digital flexor tendon (DDFT) and superficial digital flexor tendon (SDFT) pass through. In this canal, the digital flexor sheath is lined with synovial membrane, and when all is well, the DDFT and SDFT can glide smoothly through. However, constriction of the flexor tendons can occur, causing pain and lameness. Constriction can be caused by thickening/fibrosis of the PAL (also known as desmitis), enlargement or tendonitis in the SDFT or DDFT, or inflammation of the digital sheath's synovial membrane.

Constriction can be caused by direct or indirect trauma in the area, and, as with other tendon and ligament issues, damage can occur slowly or suddenly. If the tendons are not significantly damaged, prognosis is usually good. PAL constriction is often resolved with annular ligament desmotomy (surgical incisions), which we'll discuss on page 126.

Curb

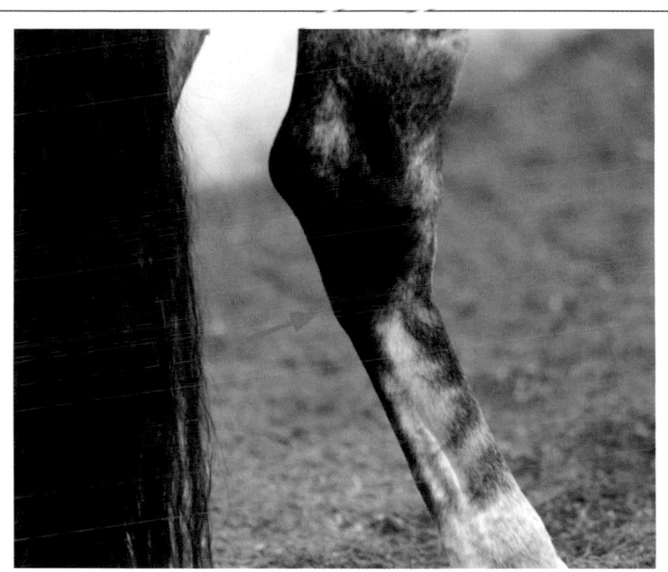

- The historic term *curb* refers to swelling in the lower back part of the hock and thickening of the long plantar ligament.
- While the term *curb* can accurately describe the inflammation that occurs, it doesn't accurately define causes.
- Typical curb can be caused by injury to the SDFT, the DDFT, the long plantar ligament, or the tissues around them.
- Your veterinarian should examine any inflammation to determine the cause.

Intercarpal Ligament Tearing

- The carpus or knee has seven small bones, and the ligaments that attach within the knee can become torn.
- Medial palmar intercarpal ligament tearing refers to ligament tearing in the middle carpal joint, but this is not the only knee joint where tearing can occur.
- Intercarpal ligament tearing is often accompanied by synovitis.
- Many horses with tearing to the ligaments within the knee have bone chips that can be removed via arthroscopic surgery.

TENOSYNOVITIS

Tenosynovitis is inflammation of the tendon sheath's synovial membrane and can appear as thoroughpin or windpuffs

Friction is an enemy of tendons. To do their job, tendons need to glide smoothly. Where tendons pass over joints or in places where they change direction, they are enclosed in tendon sheaths lined with synovial membranes. Tenosynovitis is inflammation of the tendon sheath's synovial membrane and often the fibrous layer of tendon sheath as well. Effusion is an accumulation of fluid, and tenosynovitis is usually visible due to synovial effusion and distension of the tendon sheath. Tenosynovitis is divided into different types.

Idiopathic tenosynovitis is often considered just a blemish because it is synovial effusion without any associated inflammation, pain, or lameness. It most often affects the extensor

Thoroughpin

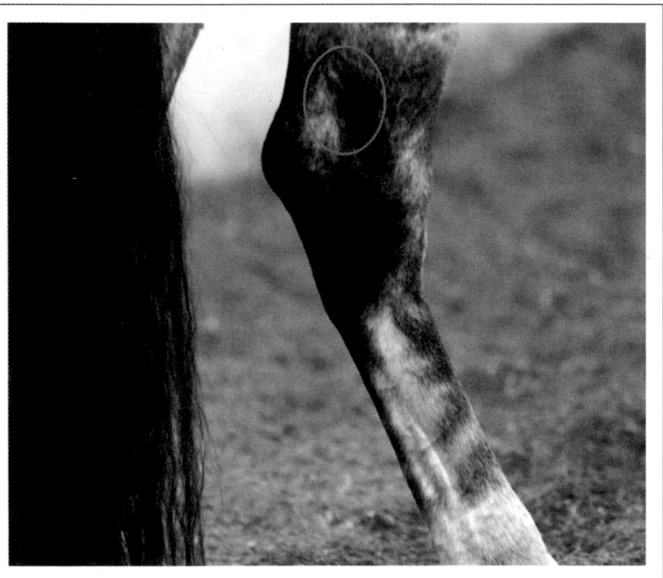

- When tenosynovitis affects the tarsal sheath of the DDFT just above the hock, it is called thoroughpin.
- With thoroughpin, distention of the sheath can usually be seen on both sides of the leg.
- When it's not associated with pain or lameness, thoroughpin is mainly a blemish.
- However, the swelling must be differentiated from bog spavin (see page 94) or other conditions, so it's important to have the vet examine any swellings.

Windpuffs

- The DDFT sheath is one of those most commonly affected by tenosynovitis.
- The terms *windpuff* and *windgall* are often used to describe tenosynovitis (and its accompanying synovial effusion) in the DDFT sheath near the fetlock.
- However, tenosynovitis in the fetlock area can be confused with synovitis of the fetlock joint (sometimes also called windpuffs), so the vet will need to examine the horse for an accurate diagnosis and treatment plan.

tendon over the horse's knee or the deep digital flexor tendon (DDFT), such as where it's enclosed by the tarsal sheath just above the hock (where it's known as thoroughpin). Because there is no pain or lameness in idiopathic tenosynovitis, treatment isn't necessary except to reduce the blemish.

Acute tenosynovitis shows up quickly and is associated with pain, heat, and often lameness. The vet will likely use ultrasound to investigate the cause, which can be direct trauma to the area, such as a knee hitting a jump, or tendonitis (injury and inflammation in a tendon), or some other type of internal friction or pressure. Prompt treatment usually has a good outcome, but waiting can lead to chronic tenosynovitis.

The last type of tenosynovitis is septic (infectious) and is associated with heat, pain, swelling, and often severe lameness. Septic tenosynovitis can be caused by a puncture wound or laceration, or when an infection spreads to the area. The vet will confirm the diagnosis with synovial fluid analysis and then drain and irrigate the area as well as prescribe antibiotics. Septic tenosynovitis can be deadly, but if treatment is prompt and successful, the horse may make a full recovery.

Acute Tenosynovitis Treatment

- Treatment for acute tenosynovitis will likely include icing as well as rest.
- Depending on the cause, treatment may also include removal of the fluid and injection of corticosteroids.
- Waiting to treat acute tenosynovitis can lead to chronic tenosynovitis, which is persistent synovial effusion and fibrous thickening of the tendon sheath, leading to compromised function, such as a knee that won't flex properly.
- In some chronic tenosynovitis, surgery may be needed.

Idiopathic Tenosynovitis Treatment

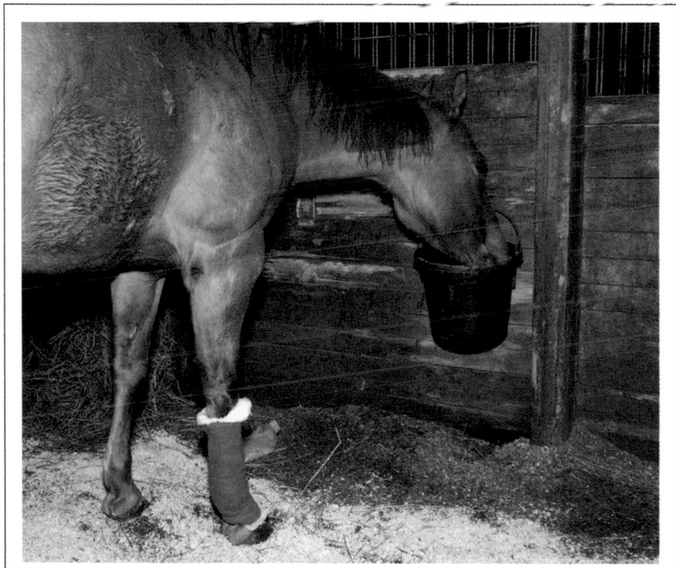

- Idiopathic tenosynovitis is not associated with pain or lameness, so it's mainly considered a blemish, and treatment is optional.
- Sweat wraps sometimes help decrease the distension associated with idiopathic tenosynovitis.
- The vet may recommend withdrawing the fluid and injecting corticosteroids or hyaluronic acid. However, the effusion may return.
- If lameness appears in the affected leg, have the vet reevaluate the horse.

LIGAMENT AND TENDON SPLITTING

Surgical methods can be beneficial to certain tendon and ligament injuries

Surgical procedures, including tendon splitting and ligament desmotomy and desmoplasty, can have favorable outcomes when treating certain tendon and ligament issues. There are several possible procedures. Many of them can now be performed with ultrasound guiding them and the horse standing up, but some may require full anesthesia.

Tendon splitting involves surgical incisions made into the tendon lesion (where tearing occurred). Tendons don't have a lot of blood vessels, and tendon splitting seems to help vascularization, improve blood flow, and speed healing. The superficial digital flexor tendon (SDFT) is the most commonly injured tendon, and certain SDFT injuries are good candidates

Suspensory Desmoplasty

- Suspensory desmoplasty is similar to tendon splitting, with incisions made through the lesions.
- The suspensory ligament is the most commonly injured ligament, and suspensory desmoplasty may be recommended for horses with chronic lesions that did not heal after rest.
- Healing after the surgery takes approximately three to six months, and results are usually favorable.

Constricted Palmar Annular Ligament

- This Thoroughbred show jumper was diagnosed with a constricted palmar annular ligament on her right front leg, and her vet recommended the surgical procedure of an annular ligament desmotomy.
- Her recovery included nine weeks of gradual rehabilitation, from short hand walks to walking under saddle, then walk/trot and finally brief work at all gaits.
- She is able to return to full work, including jumping, although the owner is careful to monitor the horse's soundness.

for tendon splitting. Early treatment can decrease the size of the lesion and improve tissue repair. Most horses treated are able to resume their previous athletic work after healing.

Another procedure, desmotomy (or transection) of the superior check ligament can sometimes help chronic cases of SDFT tendinitis. However, the procedure is not always successful, and certain horses are re-injured or get suspensory ligament desmitis following superior check ligament desmotomy.

Palmar-plantar annular ligament (PAL) constriction can also cause problems for the SDFT and deep digital flexor tendon (DDFT), in which case, PAL desmotomy may be recommended. This procedure transects the ligament to relieve constriction.

Desmotomy can also be performed on other ligaments for varying reasons. For example, standing ligament splitting can be done on horses suffering from upward fixation of the patella; however, synovitis and osteoarthritis may result.

Tendon Splitting

- Tendons identified via ultrasound as having large core lesions are often good candidates for tendon splitting.
- Tendon splitting involves making an incision with a small scalpel through the skin, tendon, and into the core lesion.
- The procedure helps drain the excess fluid/edema in the area and improves vascularization.
- Studies show most treated horses benefit from a reduced lesion size, injury grade, and tendon diameter.

Recovery

- Although tendon splitting produces a positive outcome in most cases, as with most surgical procedures, a carefully managed recovery period is essential.
- Six months of restricted activity is usually warranted.
- Stall rest with hand walking is often prescribed for the first month and sometimes longer.
- Follow up ultrasounds will influence how quickly the horse is returned to riding, with the intensity of under saddle work increased slowly over time.

UPPER LEG FRACTURES

Learn important general fracture terminology plus common fractures of the upper leg

First, let's bone up on general fracture terminology: comminuted fractures are fragmented and are more severe than simple ones; displaced fractures are where the bone is knocked out of place and are more severe than nondisplaced; complete fracture means the fracture goes all the way through the bone and is more severe than incomplete or stress fractures. Fractures can be open (compound), meaning the skin is broken and the bone may be protruding, or closed. Fractures can also be articular (involving a joint surface), which is often more severe than nonarticular. In addition, fractures can occur in various directions.

Fractures to the horse's radius (front thigh bone) and the

Radius and Ulna

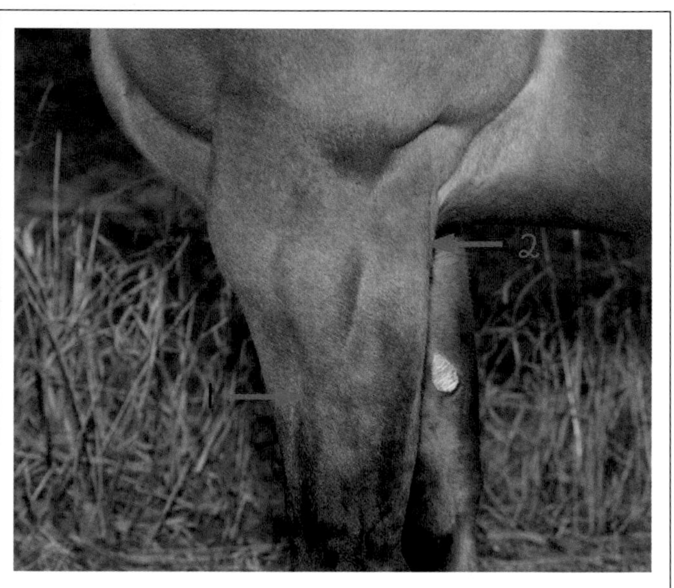

- The radius (arrow 1) is the horse's front "forearm" bone, and the ulna (arrow 2) is a smaller bone in the elbow area, as shown in the photograph above of a healthy horse.
- Kicks and accidents are the most common cause for fractures to the ulna and radius.
- A wound at the site of injury may be present.
- Fractures to the ulna or radius cause swelling in the area and severe, often non-weight bearing, lameness.

Radius Fracture

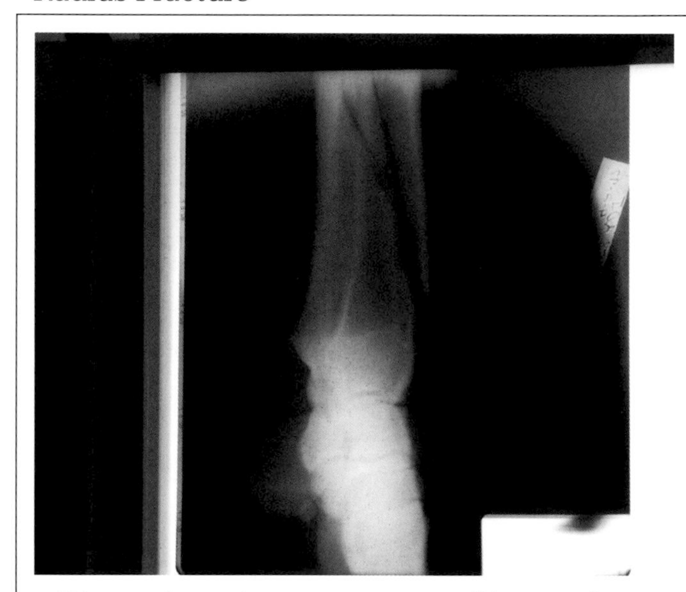

- This x-ray shows a horse with a fractured radius.
- Fractures to the radius are often caused by kicks or other accidents.
- Horses with non-displaced fractures may be able to walk with some weight on the limb. Displaced fractures will be more obvious, as the horse will not bear weight on the limb.
- Swelling and soreness around the fracture and a wound near the site of impact may be present.

ulna above it in the elbow area are relatively common. Treatment for fractures of the radius vary depending on the fracture, ranging from cast alone or cast with transfixion pins, to surgery with internal plate fixation. If surgery is required, coming out of anesthesia in a pool or sling can help avoid re-injury. Prognosis for displaced fractures in adult horses is poor. Postsurgical complications include infection or laminitis in the nonaffected limb. The prognosis for nondisplaced fractures is more favorable.

RED LIGHT

If you suspect a fracture, it's very important to keep the horse still until the vet arrives. If the horse tries to put weight on the leg or move around, he could cause severe soft tissue damage or turn an incomplete fracture into a complete fracture, or create a compound or open fracture.

Splinting and Treatment

- Fractures to long bones such as the radius need immediate splinting, but unless the owner is trained on splint application, it's best to let the vet splint the leg.
- Fractures to the ulna in adult horses are often articular, complete, and somewhat displaced. Minor fractures may be treated with stall rest and splinting, while more serious fractures often need internal fixation with plates. With appropriate treatment, prognosis is good.

Stifle and Patella Fractures

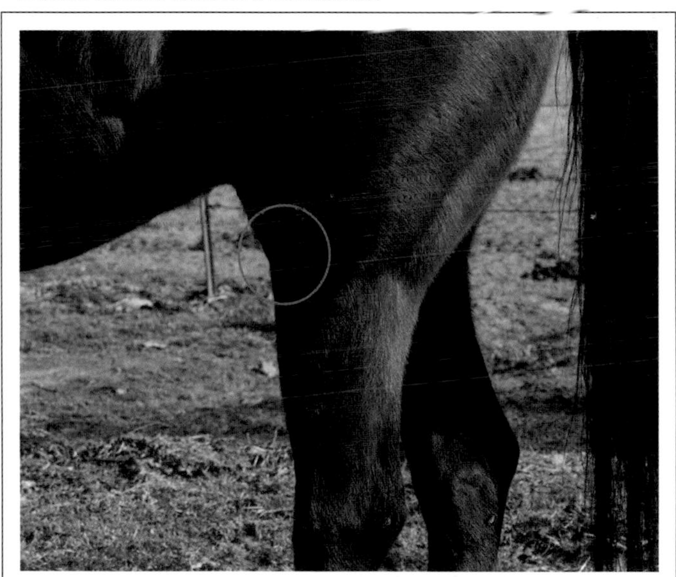

- The stifle and patella are shown in the photograph above of a healthy horse, and fractures to these bones are rare.
- Patella fractures can be associated with soft tissue trauma involving ligaments or the stifle joint. Fractures to the patella most often occur from direct trauma such as a kick or hit to a jump.
- If the fracture is not displaced, the horse may recover with rest.
- Displaced fractures of the patella may need surgery with internal fixation.

KNEE AND HOCK FRACTURES

The carpus and tarsus are made up of several small, stacked bones that can fracture

The horse's carpus (knee) is made up of seven small bones stacked together with three notable joints. Intra-articular (within the joint) fractures most commonly occur in the carpus and include chip fractures, slab fractures, and/or comminuted fractures.

Intra-articular carpal fractures can be caused by direct trauma or repeated concussion and commonly affect racehorses and other performance horses. Fast speeds, fatigue, poor shoeing, poor footing, and poor conformation can all contribute. Heat, pain, and joint distention are some of the signs of possible intra-articular carpal fractures. Treatment varies from stall rest to surgical stabilization. Comminuted (frag-

Accessory Carpal Bone Fracture

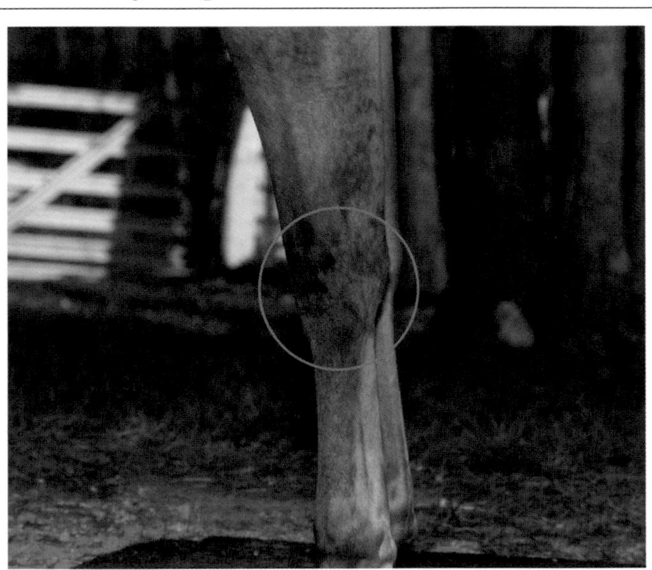

- The carpus is made up of many small bones stacked together with the accessory carpal bone sticking out the upper back.
- Fractures to the accessory carpal bone are commonly seen in racehorses, eventers, and hunters/jumpers. (Healthy horse shown here.)
- Horses suffering from accessory carpal bone fractures will likely show lameness and may have swelling or distention of the carpal sheath.
- Treatments vary from rest to surgery. With proper treatment, the prognosis is good for a return to work.

Bone Chips

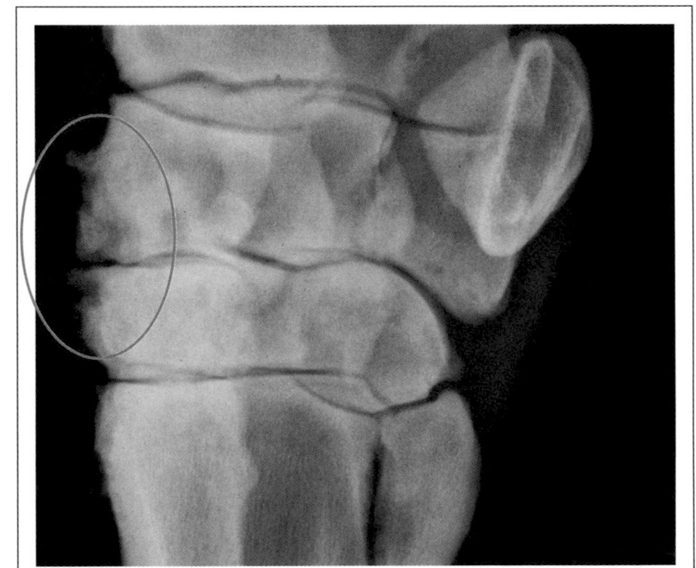

- This x-ray shows chip fragments in the carpus (knee).
- Bone chips or osteochondral fragments most commonly occur in the carpus and fetlock.
- Bone chips can occur as a result of trauma, when the bones and joints are unevenly loaded, or as a result of developmental issues such as osteochondrosis.
- If the bone chip is causing problems, it is commonly removed via arthroscopic surgery.
- This horse developed secondary DJD and was retired.

mented) fractures are usually the most difficult to treat and can have a poor prognosis. The tarsus (hock) is made up of six small bones. Intra-articular fractures of the tarsocrural joint can occur but are uncommon. Slab or sagittal fractures of the central and third tarsal bones are most commonly seen in racehorses (Standardbreds especially). The amount of swelling and lameness and whether joint effusion is present will depend on which bone is affected. Some fractures can be treated with rest alone, while others need surgery with screw fixation or other treatment.

ZOOM

Chip fractures (osteochondral fragments) are fragments of bone and cartilage that break off. If the bone chip is covered in scar tissue or in a low-motion area, it may not cause a problem. However, if the chip rubs against bone, it sheds debris that irritates the joint and can cause pain, lameness, and eventually degenerative joint disease (DJD) and osteoarthritis.

Carpal Sagital Fracture

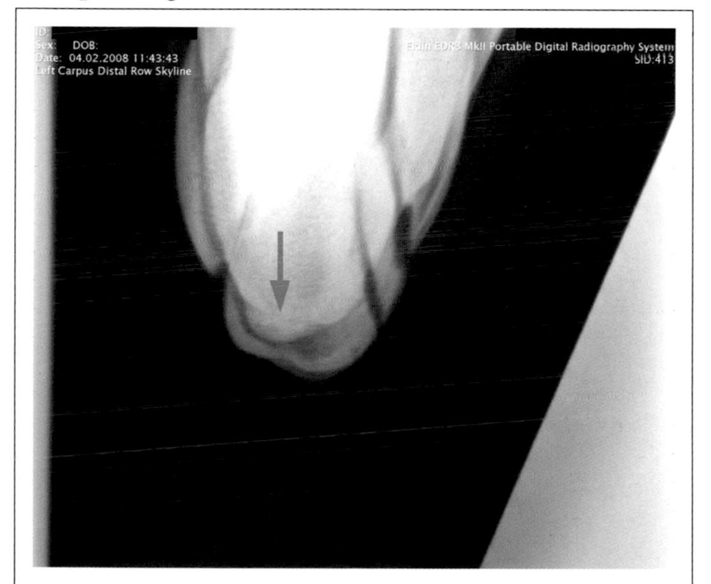

- This photo shows a carpal sagital fracture that occurred in a mare during turnout.
- Surgery was conducted for debridement, but due to the site a screw could not be placed.
- The mare was treated with stall rest, non-steroidal anti-inflammatory medication, and extracorporeal shockwave therapy.
- One year later it is still present via x-ray and has healed as a non-union fracture of the third carpal bone. The mare is to be retired as a broodmare.

Hock/Tarsus Fracture

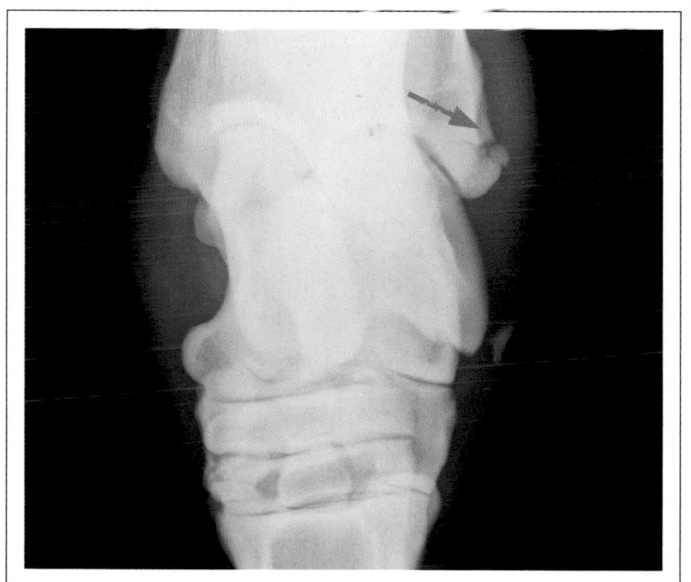

- This x-ray shows a lateral malleolus fracture of the hock.
- The malleolus is a small prominence on the distal (lower) end of the tibia (the long bone above the hock).
- The fragment was screwed back into place, and the horse is now sound under saddle.
- X-ray is used to diagnose most fractures, but in some cases intrasynovial anesthesia or nuclear scintigraphy may also be employed to help accurately pinpoint fractures of the tarsus or carpus.

CANNON AND SPLINT FRACTURES

The cannon bone or splint bones can fracture due to injury and may need surgery

The cannon bone (also called the third metacarpal in front or third metatarsal in back) and the splint bones (also known as second and fourth metacarpal in front and second and fourth metatarsal in back) can fracture anywhere along their length. Fracture-causing injuries may also involve other bones, soft tissues including tendons and ligaments, or the fetlock or knee joints.

About half of cannon bone fractures are compound or open fractures (skin is broken and bone may be protruding). Most cannon bone fractures occur from traumatic event injuries, such as kicks from other horses or run-ins with fencing or holes. A small stress fracture (incomplete fracture) can also

Cannon and Splint Bones

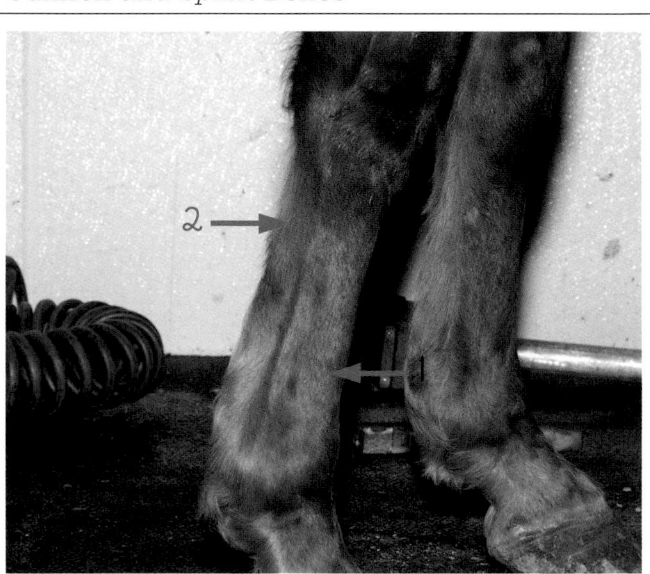

- The cannon bone (arrow 1) is the main bone between the fetlock and the knee. Each cannon bone has two splint bones—one on either side (arrow 2).
- Fractures to splint bones often occur in the lower part of the splint bone, where the bone is thinner.
- This lower part of the splint bone can be surgically removed if the injury causes long-term problems.
- A fractured splint bone is usually accompanied by heat, pain, and swelling. (Healthy horse shown here.)

Splint Fracture

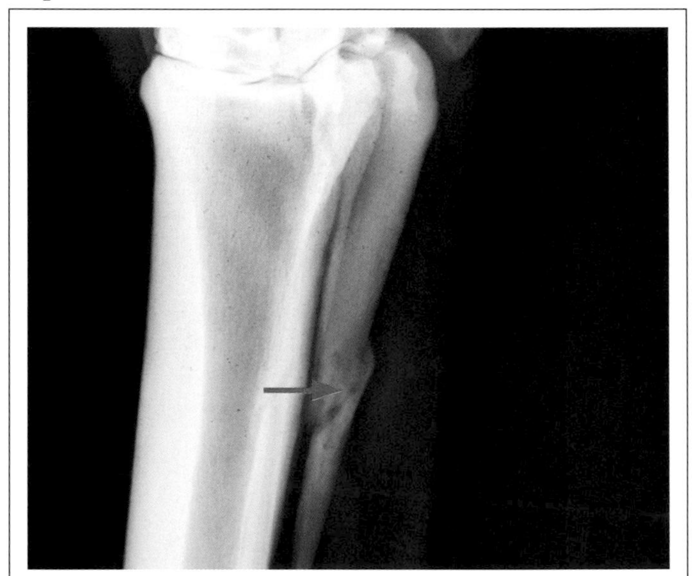

- This x-ray shows a splint bone fracture.
- X-rays will diagnose a splint bone fracture and differentiate it from a splint.
- Splint bone fractures are often caused by kicks or interference injuries.
- When splint fractures are caused by external trauma, there will often be an accompanying wound.
- This horse injured himself kicking a fence in turnout. After rest and physical therapy, he returned to full training.

spread, leading to complete cannon bone fracture.

After a fracture, there is usually pain, heat, and swelling in the area. The amount of swelling and lameness depends on the severity and location of the fracture. A wound at the impact site may also be visible.

Suspected fractures are a veterinary emergency. X-rays are the frontline of fracture diagnosis, and usually multiple views are taken. Treatment for cannon bone fractures and the outcome depend on the severity of the break, its location, whether it's an open or closed fracture, and the horse itself.

RED LIGHT

If the horse must be transported to the hospital after a fracture, the vet will likely splint or cast the leg for stability first. If available, a trailer sling can also be used for certain fractures. Trailers with ramps are best for injured horses, and horses with front leg fractures do best riding backward.

Recovery Pools

- Horses waking up from major surgery to repair a fracture often do best if they can be placed in a recovery pool, which many major equine hospitals now have.
- In a warm recovery pool, horses can be upright without placing weight on the limbs, and if they flail, they don't hurt themselves.
- The horse may be lowered via a harness into a large raft with rubber sleeves for his legs.
- Once the horse is fully awake, he's moved from the pool to a stall.

Cannon Bone Fracture Treatment

- The most common surgical treatment for cannon bone fractures involves placing two dynamic compression plates along the length of the bone and securing them in place with screws.
- A cast followed by six months of stall rest is usually necessary after surgery.
- Complications after surgery can include infection, or laminitis in the noninjured leg.
- Depending on the horse's intended use, the plates may later be removed or they can be left in.

PASTERN FRACTURES

Several types of fractures can affect the long and short pastern bones and the proximal sesamoid bones

Both the short pastern (second phalanx) and long pastern (first phalanx) can be fractured in a number of ways and places. Chip fractures can affect the pasterns as well. Pastern fractures can occur in front or hind legs, but short pastern bone fractures are most common in the hind legs.

Western performance horses, seem to be especially prone to hind leg pastern fractures, including comminuted long pastern bone fractures. This is attributed to the compression and twisting created during the sudden stops and sharp turns these horses perform. Racehorses are also prone to pastern fractures, and hunters and jumpers sometimes suffer from noncomminuted long pastern bone fractures.

Pasterns and Proximal Sesamoid Bones

- The short pastern (arrow 1) sits halfway in and halfway out of the hoof. Above it is the pastern joint and then the long pastern (arrow 2).
- Above the long pastern is the fetlock joint (arrow 3), and at the back of the fetlock sit the proximal sesamoid bones (arrow 4).
- Fractures to the proximal sesamoid bones are common in racehorses.
- Proximal sesamoid fractures most often occur from weight-bearing forces exerted on the bone by the suspensory ligament and distal sesamoid ligaments.

Pastern Fracture

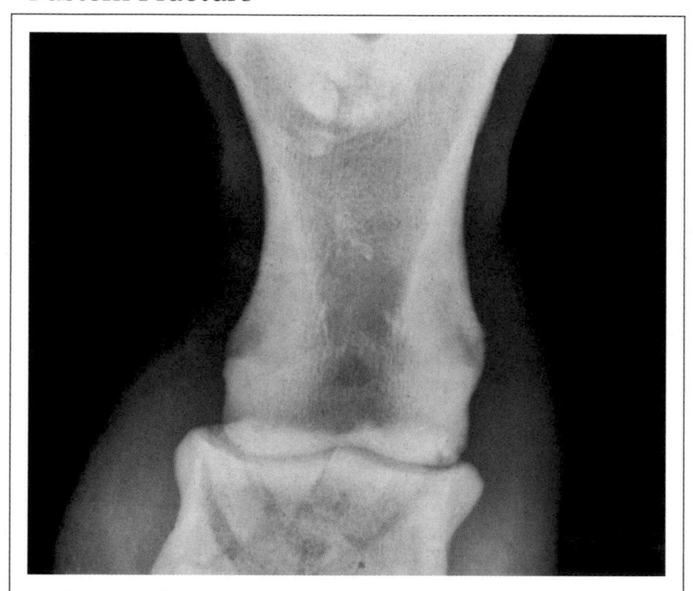

- This x-ray shows a severe fracture that shattered the short (or first) pastern bone of a nineteen-year-old horse during lunging.
- While severe fractures like the one shown here would likely produce non-weight-bearing lameness, less severe fractures may produce more moderate lameness and be noticeable from swelling above the coronary band/coronet and pain on flexion.
- Traction devices such as shoe studs can make a horse more susceptible to pastern fractures.

Pastern fractures are rarely open, but can be caused by direct trauma or a penetrating wound. Fractures may occur when the horse is at play or when he's performing.

Fractures that are simple and nonarticular will usually produce less lameness and swelling than comminuted or articular fractures. Prognosis and treatments vary greatly depending on the type of fracture and its location. A horse that heals well after a minor pastern fracture may return to work. Severe pastern fractures can have a poor prognosis for usability or even survival.

RED LIGHT

Cost for treating serious fractures can be extremely high, so discuss expenses up front. Before deciding on fracture treatment for a particular horse, the vet and horse owner should consider the horse's size, temperament and current condition, and the location and severity of the fracture. All these factors will influence the outcome of treatment.

Pastern Fracture Treatments

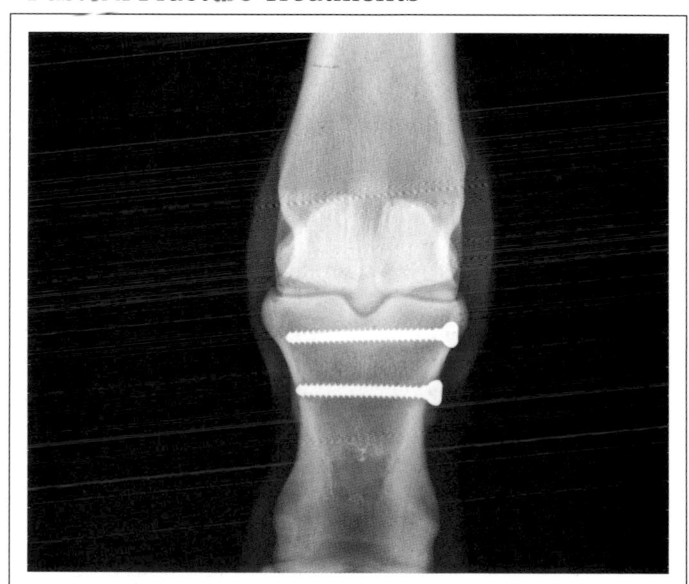

- Some more minor pastern fractures can be treated without surgery by using a combination of rest and casts or pressure bandages.
- Other pastern fractures require surgery to place screws and possibly plates for stabilization.
- In this x-ray you can see that screws were used to repair a fracture to the long (or first) pastern bone.
- The screws were later removed, and the horse is now sound and able to return to riding.

Proximal Sesamoid Fracture

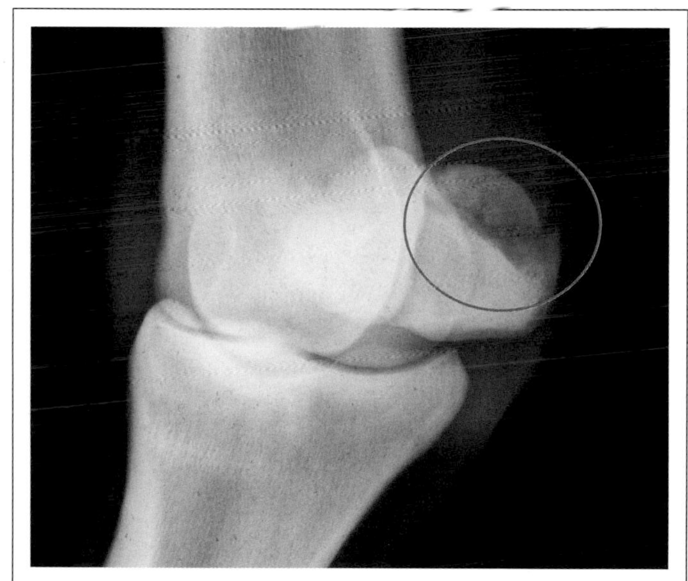

- This x-ray shows an apical proximal sesamoid fracture. The horse retired from his racing career but healed with conservative treatment.
- The word *apical* means the tip or top—the apex.
- In one study of racehorses, apical proximal sesamoid fractures most commonly occurred on the hindlimbs, and treated hindlimb fractures of this type had a better outcome than forelimbs.

COFFIN AND NAVICULAR FRACTURE

Although not extremely common, fractures to the navicular and coffin bones in the hoof do occur

While the bones within the hoof are less likely to fracture due to a kick from another horse, as often affects other leg bones, the coffin (pedal bone or distal phalanx) and the navicular (distal sesamoid) bone receive their fair share of concussion.

Coffin bone fractures usually produce sudden, severe lameness, sometimes when the horse is in the middle of a workout. Although coffin bone fractures can occur in any breed, like many other fractures, they have a higher incidence in racehorses. Horses that race on hard ground may be more susceptible.

Treatment and prognosis depend on the type of fracture and whether it's articular and affects the coffin joint or not.

Coffin and Navicular Bone

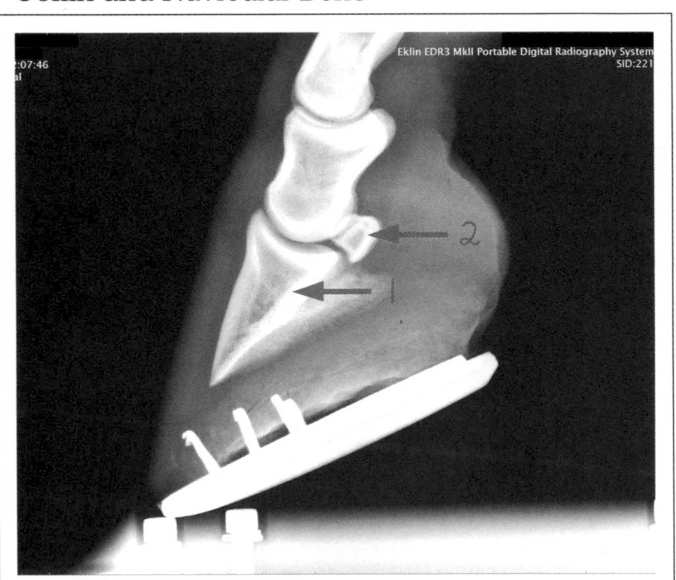

- The coffin bone, a small hoof-shaped bone, sits in the front of the hoof capsule (arrow 1).
- The navicular bone is even smaller than the coffin bone and sits below the short pastern and behind the coffin bone (arrow 2).
- The deep digital flexor tendon (DDFT) runs behind the navicular bone and under it to the coffin bone.
- Because the navicular bone sees a lot of action, it can be hard for navicular bone fractures to heal by forming a bony union. (Healthy horse shown here.)

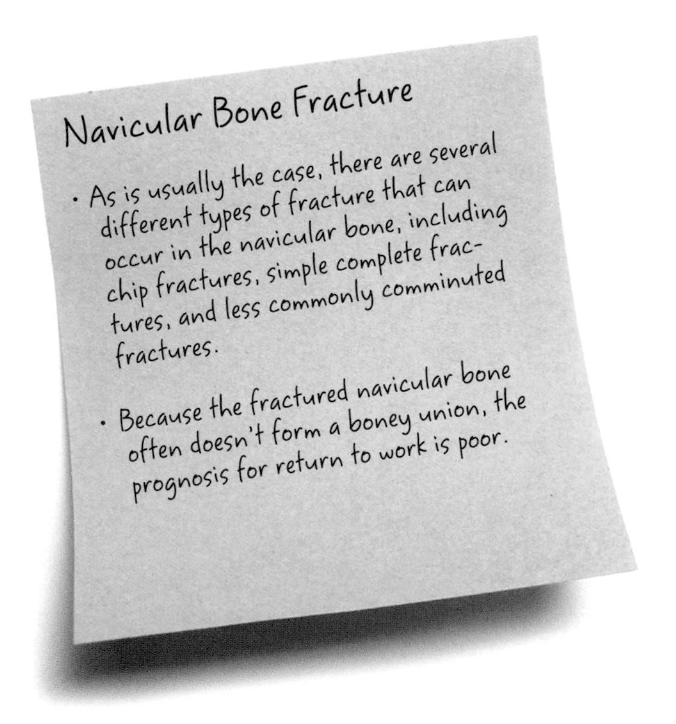

For both articular and nonarticular fractures, up to a year of rest and corrective shoeing are usually recommended. Bar shoes with quarter clips may be used. Corrective shoeing aims to immobilize the foot and limit expansion. For non-articular fractures, palmar digital neurectomy is sometimes used. Neurectomy removes a section of select nerves so that the foot doesn't feel or transmit the pain.

While the prognosis for nonarticular fractures is usually good, articular fractures have a more uncertain prognosis for future soundness. Rest and corrective shoeing may or may not work. The joint can be affected by degenerative joint disease (DJD). Adult horses with fresh articular fractures may be candidates for internal fixation with a single screw, but infection is a big risk.

Navicular fractures can be caused by concussion, trauma, or by demineralization as a result of navicular disease or sepsis from a puncture wound. As with coffin bone fractures, treatment usually consists of prolonged rest and corrective shoeing, but a bone screw may be used in certain cases.

Coffin Bone Fracture

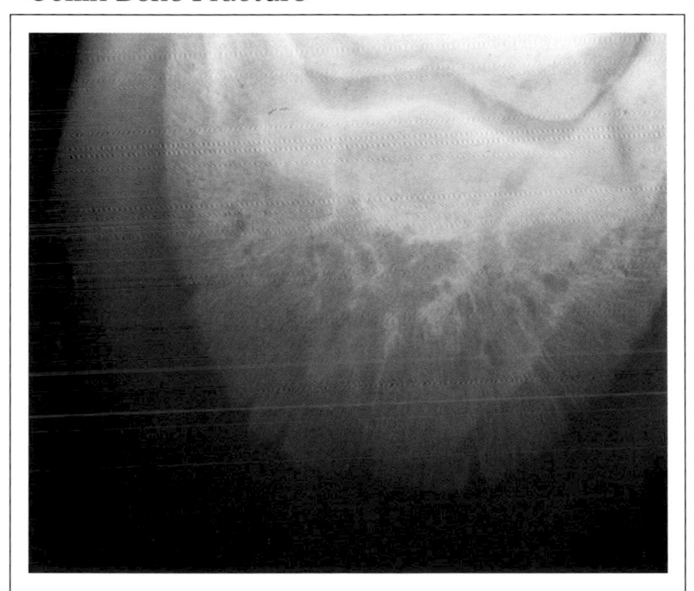

- Diagnosing a coffin bone fracture often includes nerve blocks and multiple x-ray views.
- This x-ray shows a fractured coffin bone.
- This horse was retired from riding and is pasture sound as a broodmare.
- Because of the coffin bone's unique shape, there are several different types of fractures that can occur.
- Many coffin bone fractures affect the wings of coffin bone, which are the back left and back right sides.

Hoof Testers

- Hoof testers are used to help diagnose any lameness originating from the hooves.
- Hoof testers are used to identify possible coffin or navicular bone fractures, which can be confirmed via x-ray.
- The vet will squeeze them to apply pressure to various parts of the hoof.
- Where the horse reacts to the pressure helps the vet determine where the problem comes from.

STALL REST

Injuries such as fractures require prolonged stall rest, which can be hard for horses to cope with

Most fractures require many months of stall rest, as do other severe injuries. Horses are social creatures, designed to graze and move about most of their day, so stall rest is hard on them. This is why temperament is considered when deciding on fracture treatment. A hyperactive horse may have a harder time dealing with stall rest.

If your horse has to endure prolonged stall rest, treat him as you would a friend in the hospital—try to provide him with comfort, company, and entertainment. Comfort means a safe stall, with nothing he can hurt himself on, and plenty of clean, soft bedding (and/or specialized bedding mats). If reaching his head down to eat exacerbates his injury, look for

Company

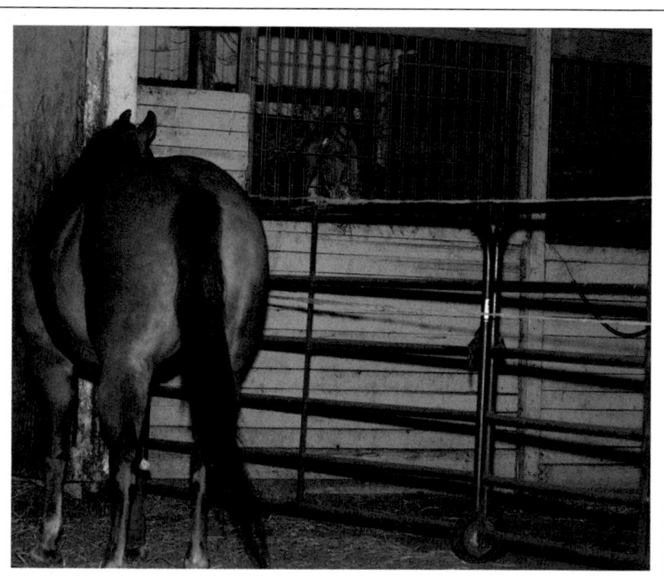

- Horses are herd animals and enjoy company.
- Injured or healing horses can't be turned out with other horses for safety reasons, but having a friend nearby is a good idea.
- Seeing and visiting with fellow horse neighbors is good for a horse's mental health and can help keep him calm and happy while he endures his prolonged stall rest.
- However, avoid situations that induce kicking at shared fencing or stalls, which can cause re-injury.

Feed

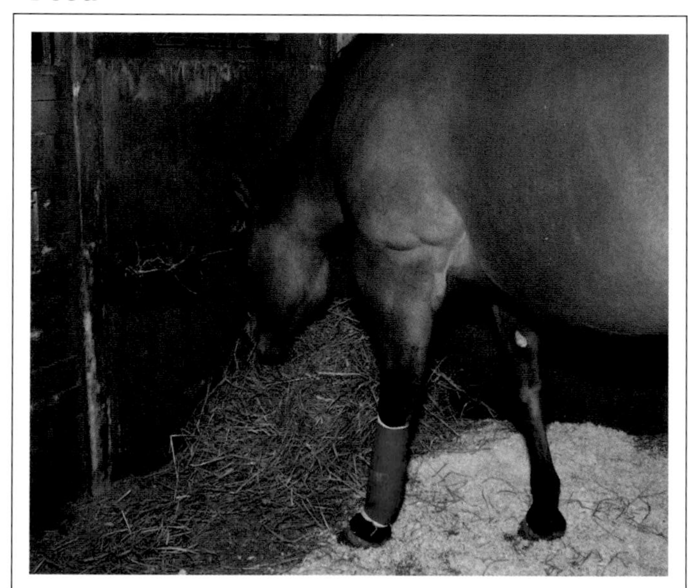

- In nature, horses spend most of their day grazing.
- A horse quickly eats concentrates and complete feeds.
- Munching on grass hay can be a good way for a horse to keep busy without gaining too much weight.
- Consult the vet to determine the best types of hay and how much to feed.

mangers or hay bags that allow him to eat with his head up.

Entertainment includes toys, eating, and observing. Talk to your vet about the size of his enclosure. Is a plus-size stall available or okay to use? Can he spend part of his day outside in a small corral? Can you let him graze on the lead line for a few minutes a day? A stall with a window lets your horse look outside and get some sunshine. You might want to move your horse to a stall where he can see more activity, such as near the arena or grooming area—provided he can watch calmly and not get agitated.

Company includes other horses nearby and humans. Spend time with your horse every day grooming him, or learn some massage techniques. Hand walking will likely be introduced during stall rest; keep in mind your horse might not act like himself! Eventually it will be time to reintroduce riding or turnout, and on the first day or two the vet may prescribe a sedative. With turnout, start small and slowly increase the size of the enclosure.

Stall Toys

- Some horses enjoy stall toys, and playing with them can help keep a horse on stall rest from getting bored.
- There are many stall toys available, including balls and toys the horses lick or spin.
- Toys may need to be rotated to keep a horse's interest.
- Keeping a horse on stall rest entertained can help avoid stall vices, such as pacing or chewing wood.

Hand Walking

- Hand walking is usually prescribed at some point in the healing process.
- The use of a stud chain may help the handler keep an energetic horse under control (but never tie a horse with a stud chain on).
- Stud chains can be looped around the horse's nose through each halter ring, or can be taken through the ring closest to the handler, over the horse's nose, and attached to one of the rings on the opposite side.

LAMINITIS

Laminitis causes failure in the laminar tissues that keep the coffin bone in place

Laminitis is a scary condition in which the laminar tissues that hold the coffin bone suspended in the hoof capsule begin to fail, or come apart (detailed in the photos below). This is painful for the horse, and the condition can progress rapidly.

While the words *founder* and *laminitis* are sometimes used interchangeably, founder actually refers to rotation or sinking of the coffin bone, which often happens in serious cases of laminitis. Laminitis most commonly affects both front feet but can affect all four feet, just the hind feet, or just one foot.

A great deal of research has been conducted trying to pinpoint what exactly takes place in the horse's body to cause laminitis. Researchers agree that problems in other parts of

Laminar Tissue

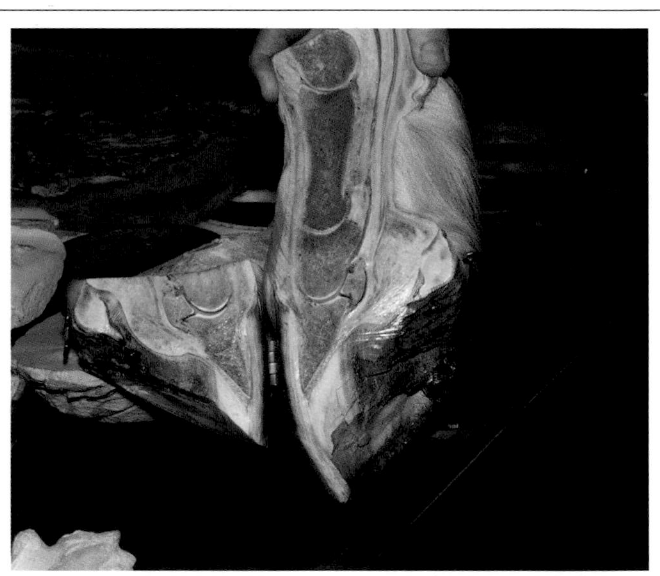

- Within the hoof, there are sensitive dermal laminae reaching out from the coffin bone to interlock with epidermal laminae from the hoof capsule.
- The dermal laminae have nerves and blood vessels and are sometimes referred to as "living" or sensitive laminae, while the epidermal laminae may be called insensitive or "non-living."
- The dermal and epidermal laminae each consist of hundreds of primary laminae and additional secondary laminae that reach out like the limbs of a tree, connecting together.

Healthy Laminae

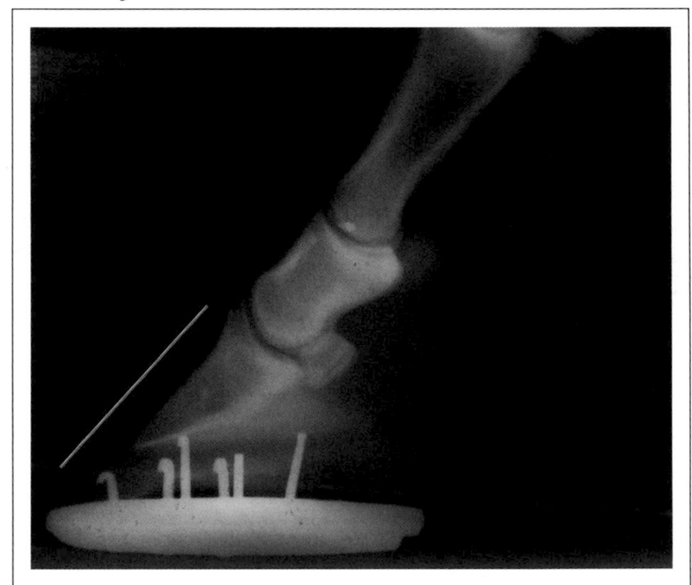

- In a healthy hoof, the primary and secondary epidermal laminae interlock with the primary and secondary dermal laminae to hold the coffin bone in place within the hoof capsule.
- In healthy hooves, these laminar tissues are very strong and virtually impossible to pull apart.
- The epidermal laminae are lined with basement membrane.
- The basement membrane is a very important sheet of connective tissue that forms the binding interface with the dermal laminae.

the horse's body often trigger laminitis, but the details are not completely clear. One theory is that laminitis starts as a vascular issue with constriction of blood flow to the foot. Other theories point to circulating toxins that damage cells or trigger enzymatic malfunction, leading to laminar breakdown. Toxins may induce matrix metalloproteinases (MMPs) production, and the MMPs damage the basement membrane.

Recent research also looks at systemic inflammatory response syndrome, which can cause multiple organ failure in humans, and in horses it may target laminar tissue and contribute to laminitis. Basically, when there's an infection, restriction of blood, trauma, or inflammation in one part of the body, it can trigger a bodily inflammatory response—the lining of the blood vessels throughout the body can become inflamed, and white blood cells can migrate into tissues and release inflammatory mediators, radicals, and MMPs, which can damage the basement membrane (the binding interface between the dermal and epidermal laminae).

Continued research will likely shed more light on exactly what takes place in the horse's body to trigger laminitis.

Laminar Breakdown

- When a horse has laminitis, the laminar tissues and basement membrane begin to fail; this can take place in certain parts or all over the hoof.
- Research shows that in a horse with laminitis, the basement membrane gets empty pockets, then becomes patchy and loses its integrity.
- The laminar tissues in horses with laminitis become thinner.
- The secondary epidermal laminae and the secondary dermal laminae begin to pull away from each other.

Founder—Rotation and Sinking

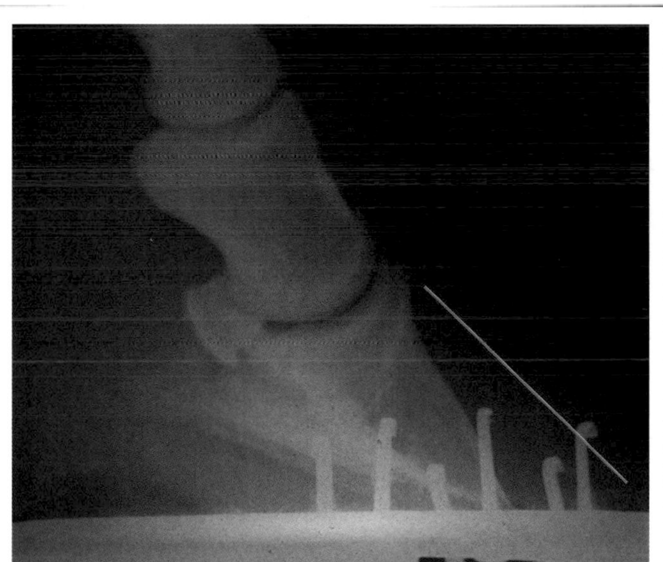

- If the laminar tissues along the front top of the coffin bone fail, then the pull of the deep digital flexor tendon (DDFT) at the back of the coffin bone may cause the front tip to rotate downward (shown here).
- When the laminar tissues fail all around, the coffin bone may lower or sink toward the sole.
- These are both serious, life-threatening conditions, and all of these events can take place within hours or days, which is why prevention and prompt treatment are key.

PREDISPOSING FACTORS: PART ONE

Carbohydrate overload and insulin resistance put horses at risk for developing laminitis

There are many things that can increase the likelihood a horse will get laminitis. Tops on that list are dietary and insulin-related issues.

In some studies, carbohydrate or starch overload caused by eating too much lush grass is the number one cause of laminitis. Laminitis can also be triggered by sudden changes in diet or an overload of grains or concentrates, such as when a horse gets into the feed room and overindulges. One possible explanation for carbohydrate overload is that an excess of quickly fermenting carbohydrates can kill the good bacteria and disrupt the pH in the large intestine, irritating the intestinal lining. Then, bad bacteria may be ab-

Cushing's Disease

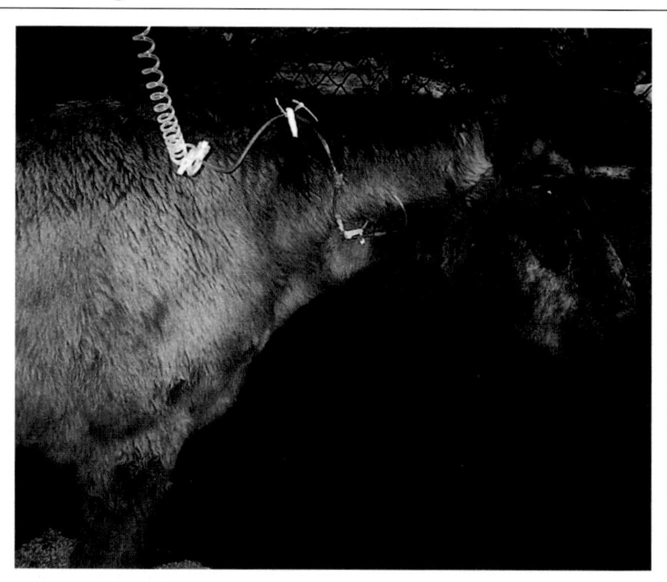

- Cushing's disease is a pituitary problem that creates hormonal imbalances.
- Older horses are especially prone to Cushing's.
- Many horses with Cushing's have a long, curly coat that never sheds out. A cresty neck as seen above is another possible sign.
- About half the horses with Cushing's will be affected by laminitis, so it's important that owners work with their vets to manage the disease.

Obesity

- Obese horses, which can include those with equine metabolic syndrome, are at a higher risk for developing laminitis.
- Obese horses place added strain on their legs and hooves.
- An overweight horse may have a crease down his back, a cresty neck, and ribs that are difficult to feel. A horse of a healthy weight will have ribs you can easily feel but not see.
- Exercise can help keep your horse fit and at a healthy weight.

sorbed into circulation, eventually triggering laminitis.

Horses with insulin resistance are also prone to laminitis. Many overweight horses, horses with Cushing's disease, and horses with equine metabolic syndrome have insulin resistance. Insulin helps the body uptake, or use, glucose (sugar). Insulin resistance creates a glucose overload that can have damaging effects.

All of this means that diet plays a big role in preventing laminitis. Laminitis can end a horse's athletic career or his life, so prevention is imperative. Many of today's horses are getting grains and concentrates that they don't need or are simply getting too much feed for their needs. Work with your veterinarian to come up with the ideal diet for your horse based on his needs and what's available in your area.

In addition, if your horse is diagnosed with Cushing's or equine metabolic syndrome, he'll likely need a modified diet. For example, horses with equine metabolic syndrome require a diet very low in sugar and starch. Dietary guidelines and medication per veterinary advice can help prevent laminitis.

Diet

- Most horses are pleasure horses and don't need the added energy and calories of grains and concentrates if they're getting good quality hay or pasture.
- An accidental overdose of grains or concentrates can also trigger laminitis.
- If you keep grains or concentrates on your property, be sure they're out of reach of horses.
- Latching storage bins shut can provide added protection should the feed room door accidentally be left open.

Grass

- Horses suddenly turned out to pasture or that consume more grass than they're used to can be at risk for laminitis.
- Horses that eat grass higher in fructan than normal, such as spring and fall grass when the ground temperatures are low, are also at a higher risk for developing laminitis.
- Some horses are especially susceptible and develop chronic grass founder.
- Monitor your horse's grass intake, and be aware of seasonal changes in pasture grass.

PREDISPOSING FACTORS: PART TWO

Illness, pregnancy, and stress can all trigger laminitis in horses

Virtually any horse under stress seems to have an increased likelihood of developing laminitis. This could be due to the body's inflammatory responses, toxins from the gastrointestinal tract, from immune system reactions, or a combination of these and possibly additional as of yet unknown factors.

Horses with gastrointestinal problems like colic or inflammatory bowel diseases, diarrhea, or colitis seem to be at greater risk. Horses with Potomac horse fever (where a certain bacteria is ingested and multiplies in the intestinal tract, causing problems) have an increased likelihood of developing laminitis, as do horses with any acute secondary infection, or inflammation or infection in the chest, like pneumonia. Stressed performance horses that may be dehydrated, exhausted, or have an electrolyte imbalance are also thought

Pregnancy

- Pregnant mares are at risk for developing laminitis.
- Although it's not known exactly why pregnant mares are predisposed to laminitis, it could be partially due to the hormonal changes that accompany pregnancy and/or extra weight.
- Mares that retain their placenta after giving birth may develop uterine inflammation and infection.
- When a mare retains her placenta, it also increases her likelihood of developing laminitis.

Injury to Other Leg

- Horses that have a severe injury, such as a fracture or major tendon or ligament injury in one leg, transfer a great deal of weight to the leg on the opposite side.
- Horses with a major injury to one leg are at great risk for developing overload laminitis in the good leg.
- The racehorse Barbaro is one famous example of this phenomenon. He fractured one leg, only to develop laminitis in the supporting leg.

to be at higher risk for developing laminitis. Pregnant mares or mares with retained placenta may develop laminitis. Horses with a severe lameness in one leg may develop laminitis in other leg. All of these cause inflammatory responses.

Of course, laminitis can strike any adult horse or pony. One of the problems with laminitis is that its devastating effects can begin before there are any noticeable symptoms. By the time you realize something is wrong, the process is well underway. Laminar tissues begin to fail within hours. Treatment, of course, is essential and can help produce a more positive outcome. The only real prevention for laminitis, however, is to treat the conditions that may predispose a horse.

If your horse has one of the predisposing factors discussed here, be sure he's treated by a vet and managed carefully. You can also talk to your vet about laminitis and see if there are any precautionary measures that can be taken, such as providing foot support, icing the foot, or administering anti-inflammatories.

Road Founder

- Stress on the hooves and their mechanisms may contribute to laminitis.
- *Road founder* is an old layman's term for horses that develop laminitis after a great deal of riding or concussion on their feet.
- Thin-soled horses may be more sensitive to concussion and at a greater risk for developing sole bruises, abscesses, or other concussion-related trauma, possibly predisposing them to laminitis.

Hoof Balance

- Horses subjected to poor trimming or shoeing practices, and horses with long toe/low heel may be at risk for developing laminitis.
- Unbalanced feet create uneven loading and unnatural forces.
- If you or your farrier notices blood in the white line, separation of the white line, or increased hoof sensitivity, treat these problems seriously.
- Maintaining balanced feet is a good preventative measure for all types of lameness.

SYMPTOMS OF LAMINITIS

It's important to recognize the signs of laminitis so that veterinary treatment can begin

Laminar changes have likely begun hours or even days before a horse becomes visibly sore and shows outward signs of laminitis. Laminitis is a veterinary emergency. Prompt treatment can help control the pain and lessen the damage.

A horse with laminitis may have increased heat in the hoof and or a bounding digital pulse (pulse in the lower leg). Horses with laminitis may look as though they're walking on eggshells. The lameness may be especially visible if the horse is walked in a circle, where he may hop.

Horses with laminitis often shift their weight from one hoof to another. If only the front hooves are affected, the horse may stand with most of his weight on his hind end, and his

Laminitis Stance

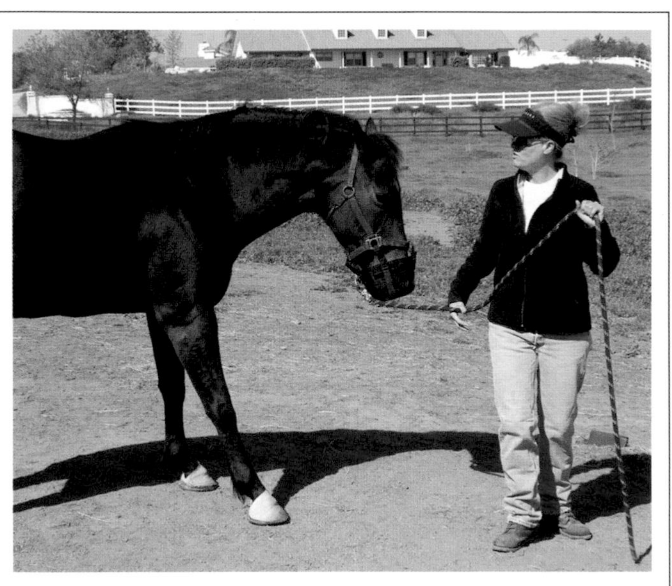

- To take weight off their front feet, horses suffering from laminitis may bring their hind legs farther underneath themselves than normal and place their front feet out in front of them farther than normal in the typical "laminitis stance."
- Horses with laminitis may also point one foot and then the other or shift their weight.
- In the photograph above, you can see this horse that is suffering from laminitis is trying to take weight off her front feet and is reluctant to move.

Digital Pulse

- Horses suffering from laminitis may have a strong, throbbing, or bounding digital pulse.
- The digital pulse can be felt in the proximal sesamoid bone area of the fetlock or midpastern (shown in the photograph above).
- When you check a horse's digital pulse, don't count beats, but rather check to see if it's significantly stronger than normal.
- Compare the digital pulse to what you have felt in the past or to your horse's other legs, such as his hind legs.

front feet farther forward than normal. If all four feet are affected, the horse may stand with front feet placed farther forward than normal and hind feet placed slightly farther back than normal.

If your horse shows signs of laminitis, call your veterinarian immediately. The vet will examine the horse and apply hoof testers to help confirm the diagnosis. Horses with laminitis often have pain in the toe region when hoof testers are applied. The vet will take x-rays to see if any signs of founder—rotation or sinking of the coffin bone—have taken place.

ZOOM

Acute laminitis refers to the early stages of laminitis, when changes in the hoof may still be taking place. Chronic laminitis more often refers to managing pain from sinking or rotation of the coffin bone, or dealing with horses that have reoccurring bouts of laminitis.

Pain

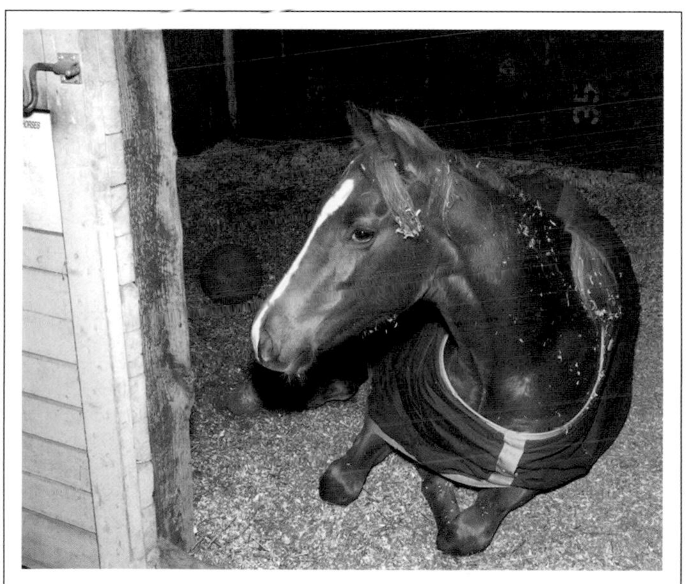

- All horses deal with pain differently.
- Some horses with pain in their feet will quietly lie down for long periods of time.
- Other horses will act agitated from the pain and may be sweaty and distressed.
- The most important thing is to know how your horse normally behaves. This way you'll recognize behavior that's out of the ordinary and may signal trouble.

Reluctance to Move

- Horses with laminitis may be reluctant to move due to the pain in their feet.
- Horses in severe pain may also stop eating or drinking.
- Anytime a horse is reluctant to move or has lost his appetite, it's important to call the veterinarian.
- The vet can give the horse an accurate diagnosis and begin prompt treatment to help avoid further damage.

TREATMENT FOR ACUTE LAMINITIS

Successfully treating the acute stage of laminitis requires treating the whole horse, not just his hooves

Because laminitis progresses so quickly, it's important to begin treatment as soon as possible, ideally within the first twelve hours. There are a number of medications that can be helpful. To try to improve blood flow to the hoof, the vet may use a vasodilator, such as acepromazine, given intravenously. The vet will also likely prescribe a nonsteroidal anti-inflammatory (NSAID) such as bute (phenylbutazone), Banamine (flunixin), or ketoprofen, and may recommend the use of DMSO as well. However, NSAIDs alone may not be enough to control the pain, so newer treatments, such as a fentanyl patch or lidocaine, given intravenously, may be added.

Icing

- Placing the horse's lower leg and hoof in ice may be helpful during the early stages of laminitis, even before pain or other symptoms are noticed.
- Cryotherapy (cold therapy) like icing may decrease laminar inflammation and slow the tissues' metabolism of harmful substances.
- The hoof and leg all the way up to the knee can be iced.
- Icing can be done using specialty boots or products, or a bucket.

Sand

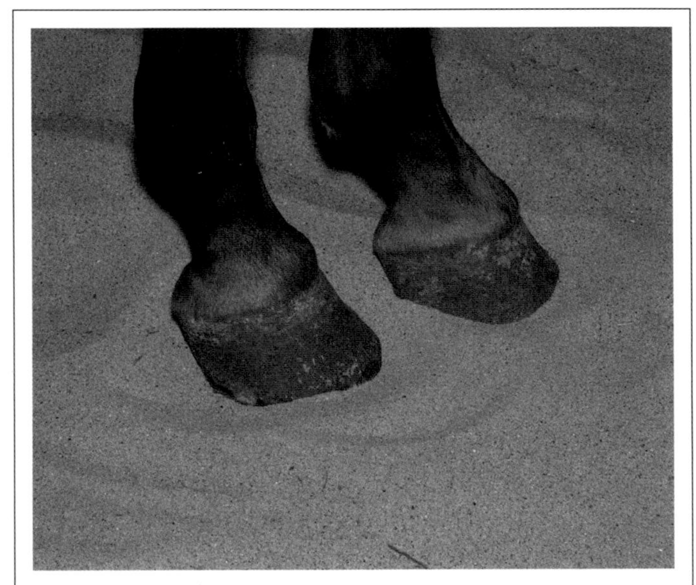

- Deep sand is one way to offer the horse's hoof added support when he's suffering from laminitis.
- Sand also gives the horse additional cushion and allows him to naturally angle his feet in a position that's most comfortable.
- When a horse develops laminitis, it's important that the vet, owner, and farrier work closely together to decide on the best hoof support and treatment for the horse.

It's also important that the veterinarian treat the primary cause of the laminitis, be it a disease or infection. A change in diet may also be recommended. Strict stall confinement is usually necessary, and the vet can advise how much if any hand walking is acceptable. It's important that the stall be well bedded to encourage the horse to lie down and get the weight off his feet.

Traditional shoes are usually carefully removed to avoid causing the horse added pain, and mechanical support is then provided. During the acute phase of laminitis, the goal is usually to give the hoof support, especially from the tip of the frog back. Products like Styrofoam that mold around the horse's frog are often used but may become compressed and need frequent changing. Inserts called lily pads are also commonly used for support.

Some horses do successfully recover from laminitis and return to their previous work. If treatment at the acute stage of laminitis is successful, then there may be no sinking or rotation of the coffin bone and no chronic stage.

X-Ray

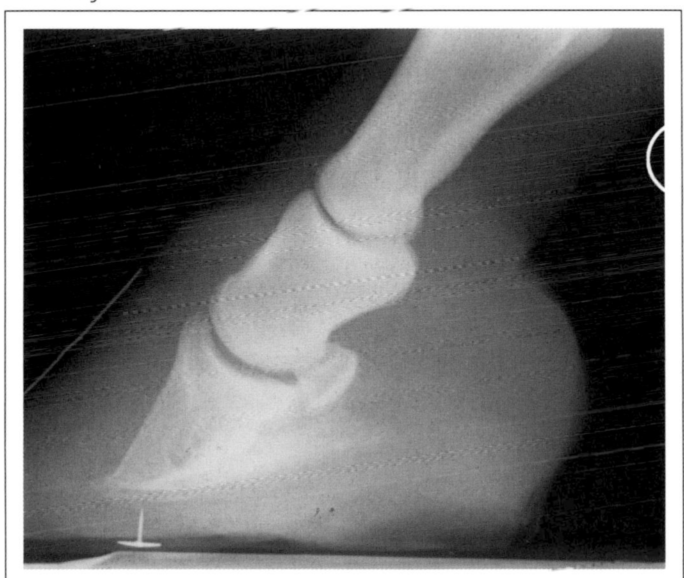

- The vet will take x-rays to see if there is any rotation or the sinking of the coffin bone.
- This x-ray shows rotation of the coffin bone, indicating the horse has foundered.
- X-rays can be taken at various points in the horse's treatment and used for comparison and to facilitate appropriate treatment.
- If changes are visible via x-ray, the x-rays can help the farrier correctly trim and support the hooves.

Shoes

- Glue-on shoes are often preferable to nail-on shoes for a horse with laminitis.
- Glue-on shoes require a less traumatic application than nail-on shoes so that putting them on won't cause the horse any additional pain.
- There are many specialty shoes and products available to give horses suffering from laminitis added hoof comfort and support.
- The veterinarian and farrier can work together to decide what's best for each individual case.

LAMINITIS COMPLICATIONS

Chronic laminitis can include sinking or rotating of the coffin bone, causing the horse constant pain

The fear when a horse develops laminitis is that he will founder, and his coffin bone will either rotate within the hoof capsule, sink, or both. Rotation and sinking are severe problems that can cause the horse considerable pain, possibly leaving euthanasia as the only humane option. In some cases, the problems can be managed, but the horse has a poor prognosis for return to work. Rotation is generally less severe than sinking because with rotation, it may mean that only the laminar tissues along the front of the hoof have failed, while tissues in the back part of the hoof may still be intact. The deep digital flexor tendon (DDFT) connects at the back of the coffin bone, and if the laminar tissues in the front

Rotation

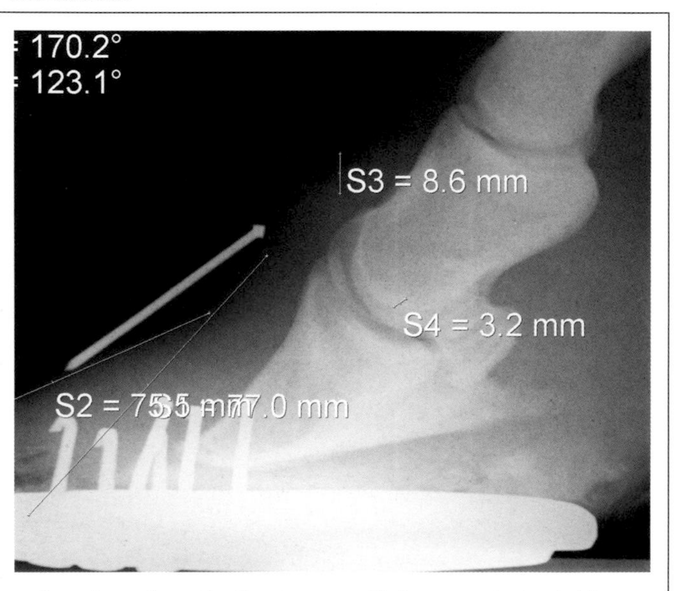

- Rotation, where the front of the coffin bone rotates down within the hoof capsule, usually occurs only in the front feet.
- This x-ray shows rotation as well as remodeling of the back underside of the coffin bone due to the laminitis and founder.
- Unfortunately, laminitis and founder can reoccur in horses and cause additional damage, which is why managing any contributing factors is always critical.

Founder and Neglect

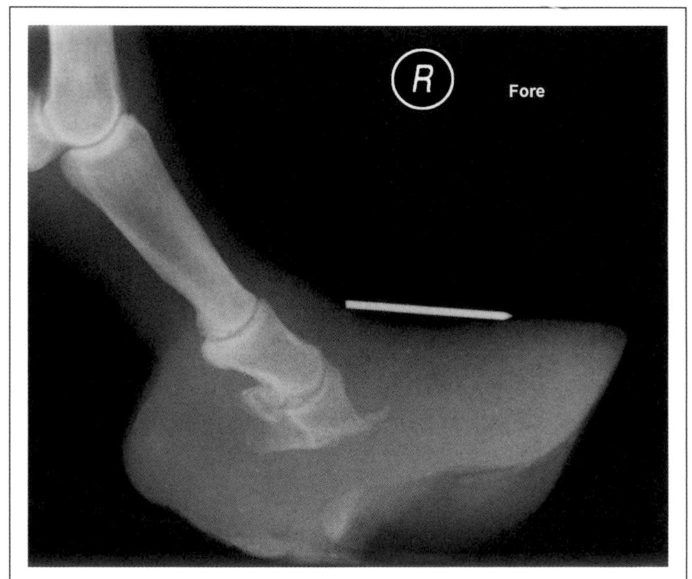

- This x-ray shows a donkey hoof—the result of neglect and severe laminitis.
- Notice the size and shape of the coffin bone; most of the coffin bone has been eroded.
- This x-ray is a good example of why it's so important to work with your veterinarian and farrier to treat and manage laminitis and founder for a more positive outcome.
- With prompt and proper treatment throughout, many horses can recover from the rotation and/or sinking caused by founder.

of the coffin bone have failed, the DDFT can pull the back of the coffin bone up, causing the toe of the bone to rotate down. The front tip of the coffin bone can then cause the horse pain at the front of his sole.

When the coffin bone sinks, it means the laminar tissue throughout the hoof failed enough that the weight of the horse shoved the entire bone down. This puts pressure on the horse's sole. The coffin bone can also sink first and then rotate. In severe cases of sinking, the bone can protrude through the sole or the entire hoof capsule can come off.

A foundered horse may develop separation at the white line and a groove above the coronary band. The horse's hoof wall also usually records the laminitis event and will eventually show ridges or a slight dishing in front.

With chronic laminitis, once any sinking or rotation has occurred, continued pain usually comes from the coffin bone pushing on the sole, rather than from the laminar tissues themselves, so providing support and protecting the sole is critical. Some chronic cases can be helped by DDFT tenotomy.

Styrofoam Pads and Inserts

- The vet may recommend Styrofoam pads be taped to the horse's hooves to provide sole support and added comfort during the acute stage of laminitis.
- Special inserts have also been designed to provide coffin bone support via the frog.
- However, it's important to only use pads and inserts under veterinary supervision, as the stage and severity of laminitis will determine what support will benefit the individual horse.

Therapeutic Shoeing

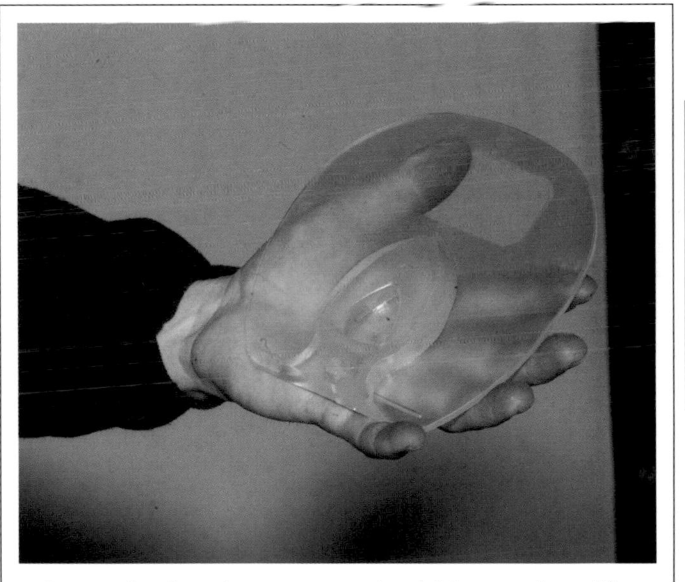

- Support for the sole and frog (one example shown above) is usually recommended for horses suffering from laminitis or founder.
- Wedge pads may be used in cases of rotation but not sinking.
- In addition to sole and frog support, egg bar, heart bar, or specialty shoes may be used for added support in place of traditional shoes.
- Breakover may be set back or the front of the shoe rockered to make movement easier for the horse.

NAVICULAR DISEASE

For accurate treatment, it's important to understand the differences between navicular disease and navicular syndrome

If you've owned or ridden horses for long, you've no doubt heard someone say, "My horse has navicular." Of course, all horses have navicular bones (also called distal sesamoid bones), so what are they talking about?

They likely mean that their horse has pain in the navicular area or rear portion of his foot. Navicular syndrome usually refers to pain in this area that has not been specifically pinpointed. Navicular disease, however, is more specific and usually refers to horses with pain stemming from the navicular bone. Horses with navicular disease have had the issue diagnosed, and X-rays, MRI, or other diagnostics have pinpointed the problem to the bone.

Navicular Bone

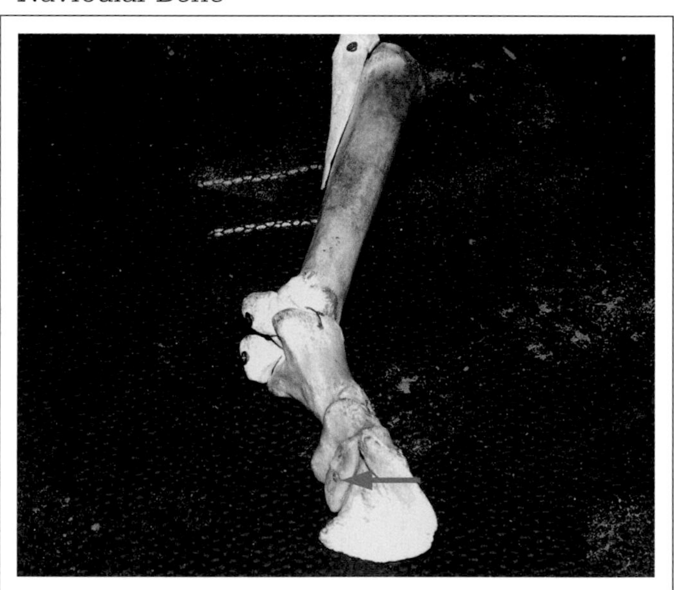

- The navicular bone (distal sesamoid) sits along the upper back side of the coffin bone and along the lower back side of the short pastern bone (seen in the photograph above with single screw in the middle).
- All three bones contact the coffin joint, which is technically called the distal interphalangeal joint.
- The navicular bone is the smallest of the three hoof bones.
- Navicular bones are longer than they are wide.

Navicular Disease

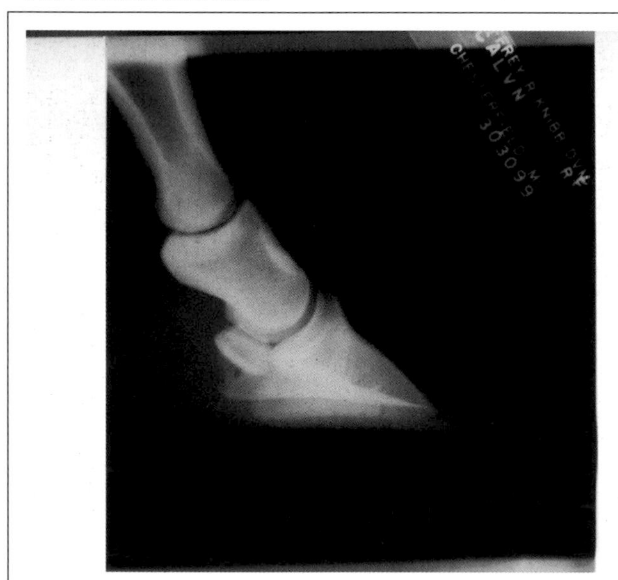

- This x-ray shows a horse with visible changes to his navicular bone.
- Horses with navicular disease often have definitive radiographic changes, unlike horses with vague navicular syndrome.
- The navicular bone naturally changes with age, but horses with navicular disease seem to show an acceleration of this process.
- Biomechanical factors—stresses placed on the bone during movement—likely play a key role in bone degeneration.

The navicular apparatus includes the navicular bursa, the collateral ligaments of the navicular bone, the distal sesamoidean impar ligaments, and the deep digital flexor tendon (DDFT). It's more common for horses to suffer from damage or injury to one of the structures surrounding the navicular bone than to have true navicular disease.

This chapter will focus on navicular disease. Horses with soft tissue injury or other causes for their foot pain need to be managed differently than horses with true navicular disease. However, keep in mind that horses can have navicular disease combined with coffin joint arthritis, or injury or damage to one of the other nearby structures.

Navicular disease is a fairly common and incurable degenerative condition that can cause a number of different changes to the navicular bone. Navicular disease produces chronic or reocurring lameness in both front feet. One front foot may show more lameness than the other, and hindlegs are rarely affected.

Research on navicular disease is ongoing, and many of the details are still unclear.

Ligaments and the Navicular Bone

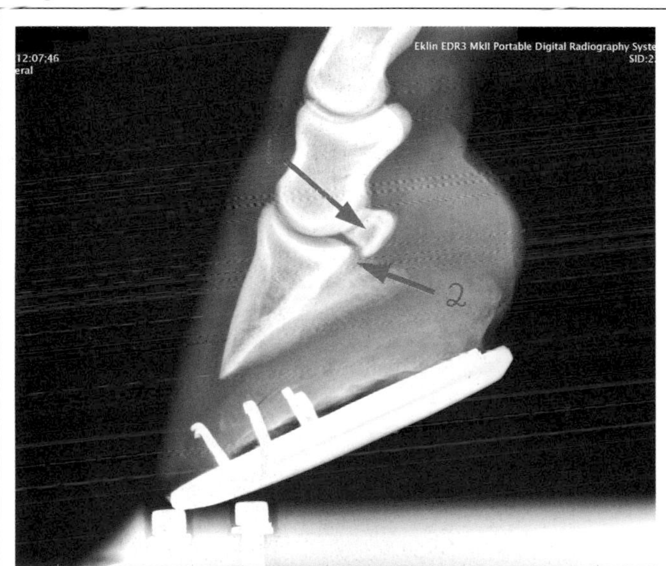

- There are several ligaments that help stabilize the navicular bone.
- The two collateral sesamoidean ligaments sweep down from the front of the leg to connect with the border of the navicular bone (arrow 1), and a branch joins the navicular bone to the cartilage of the coffin bone.
- Along the bottom of the navicular bone is the distal sesamoidean impar ligament (arrow 2), which connects with the DDFT just before the tendon meets the coffin bone.

DDFT and the Navicular Bone

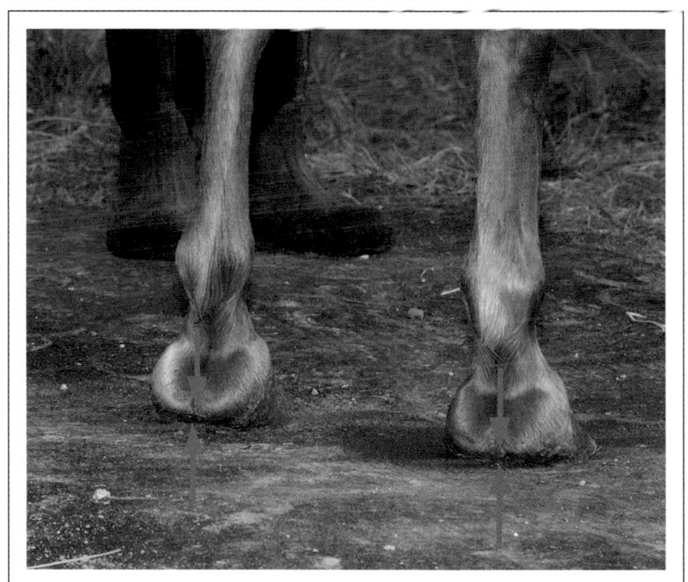

- The DDFT sweeps underneath the navicular bone to the coffin bone (as indicated by arrows).
- A fluid-filled sac called the navicular bursa lies between the navicular bone and the DDFT.
- The DDFT exerts a great deal of force on the lower section of the navicular bone.
- This relationship may cause fibrocartilage damage to the navicular bone (and DDFT) and lead to further inflammation and degeneration.

CAUSES/CONTRIBUTING FACTORS

Certain horses seem more likely to suffer from navicular disease

Although what exactly causes navicular disease is still somewhat of a mystery, there do seem to be certain predisposing factors. For example, Quarter Horses, warmbloods, and Thoroughbreds are often diagnosed with navicular disease, whereas Friesians, Arabians, and ponies are rarely affected. There may be a genetic component to the disease and/or the natural shape of the horse's navicular bone may predispose the horse to navicular disease.

Performance horses in the seven- to ten-year-old age range are most commonly diagnosed with navicular disease, although it can also be seen in younger horses just starting training, older horses, and pleasure horses.

Poor conformation likely plays a role, as well. Improper trimming or imbalanced hooves, such as upright hooves or low

Conformation and Hoof Size

- Although the causes of navicular disease are still being studied, conformation likely plays a role.
- Horses with big bodies and small feet may be predisposed to navicular disease.
- One reason for this is that the hoof has less space to distribute concussion, placing added pressure on the navicular apparatus.
- Horses with good conformation and balanced hooves are less likely to suffer from bone, joint, tendon, or ligament issues.

Obesity

- Obese horses seem to be at an increased risk for many hoof-related problems, including navicular and laminitis.
- Keep your horse fit and at a healthy weight.
- Some horses are more prone to gain weight, so work with your vet to design a feeding regimen that won't pack on added pounds.
- In addition to diet modifications, regular exercise will help your horse lose weight or ward off obesity.

heels are additional predisposing factors. Weak or crushed heels may inhibit the flexion of the coffin joint placing extra pressure on the navicular bone.

Theories have pointed to a vascular, or blood-flow problem, as the cause of navicular disease. And while some experts agree blood flow may be part of the problem, many believe it's mainly a biomechanical issue. The navicular bone sits in a precarious place next to two bones, a relatively high-motion joint and a major tendon, putting this little bone under a great deal of stress and wear.

ZOOM

Due to pain in the back of his foot, a horse suffering from navicular disease may contract his deep digital flexor muscle or land toe first to unload the heel. This also contracts the deep digital flexor tendon (DDFT), which in turn puts additional force on the navicular bone. Controlling pain and returning movement to normal is needed to stop this vicious cycle.

Work

- Working on hard ground may predispose a horse to navicular and other soundness issues.
- Horses diagnosed with navicular disease are often hard-working athletes, such as barrel racers, dressage horses, cutters, or ropers.
- Some horses with navicular disease can continue to perform with proper management, including work on more forgiving surfaces, but other horses must retire from high-performance careers.

Genetics

- Navicular disease may have a hereditary or genetic component, and certain breeds are more commonly affected than others.
- The genetically predetermined natural shape of a horse's navicular bone can influence how biomechanical forces affect it.
- Although many people retire horses with navicular disease from work and use them as broodmares, this is not advisable since they may pass the predisposition for the disease on to their offspring.

SYMPTOMS AND DIAGNOSIS

An accurate diagnosis of navicular pain is needed to create the most effective management plan

The first sign that a horse may have navicular disease is a reduction in performance. The horse may refuse his jumps, be unable or unwilling to properly extend his stride, be stiff and reluctant to perform turns, stumble, take short baby steps down hills, or show lameness in one or both front feet. In observing the horse, sometimes it looks like the pain comes from his shoulder because he changes his movement and puts more weight in his toes.

Depending on the horse and his symptoms, the vet may recommend a good farrier and start with a conservative approach before recommending expensive diagnostics, such as an MRI. For example, if the hooves are not properly bal-

Stance

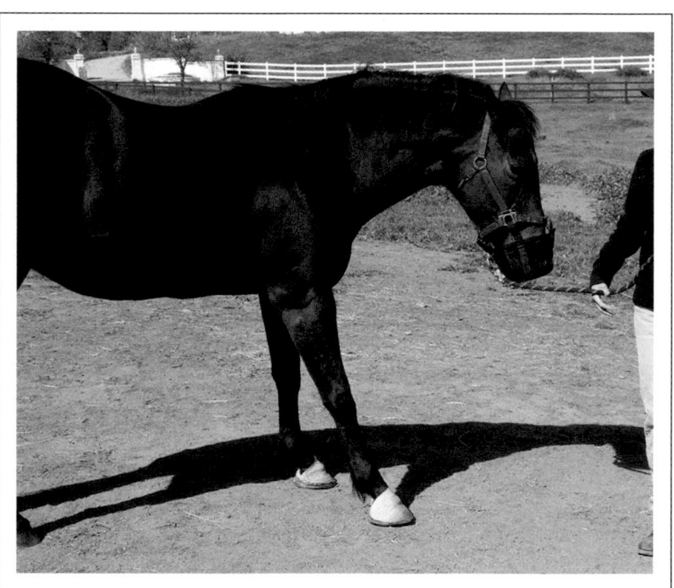

- Horses with navicular disease or navicular syndrome may stand differently, attempting to relieve the pain in the back of their foot.
- One common stance for horses with navicular pain is to "point" the most painful foot.
- If both feet are causing the horse pain, he may stand with his front feet farther in front of him than normal, similar to a horse with laminitis.

Stride

- In horses suffering from navicular disease, it may be hard to detect which front foot is lame since it usually affects both.
- One day one foot may appear lame, while the next day the other foot seems off.
- Sometimes a lameness isn't detectible per se, just a change in movement or performance.
- A short or choppy stride is common for horses with navicular disease.

anced, working to improve their balance may be the first line of attack. Some vets will wait three months before diagnosing navicular disease to see if the issues resolve.

When lameness or performance problems persist, a definitive diagnosis is needed. Hoof testers and nerve blocks will likely be used, but navicular disease can't be definitively diagnosed with these alone. X-rays are a common diagnostic tool, and many horses with navicular disease will show bony changes on x-ray, including possible cystlike lesions. Unfortunately, not all horses with navicular disease can be diagnosed via x-ray. Scintigraphy can indicate a hot spot in the navicular bone area before x-rays pick it up. Ultrasound and MRI have the added benefit of visualizing soft tissue structures as well.

While diagnostics can be expensive, it's important to treat the true cause of navicular area pain. For example, if the horse doesn't actually have navicular disease but has injury to one of the soft tissue structures, such as the deep digital flexor tendon (DDFT), he needs different treatment.

Nuclear Scintigraphy and MRI

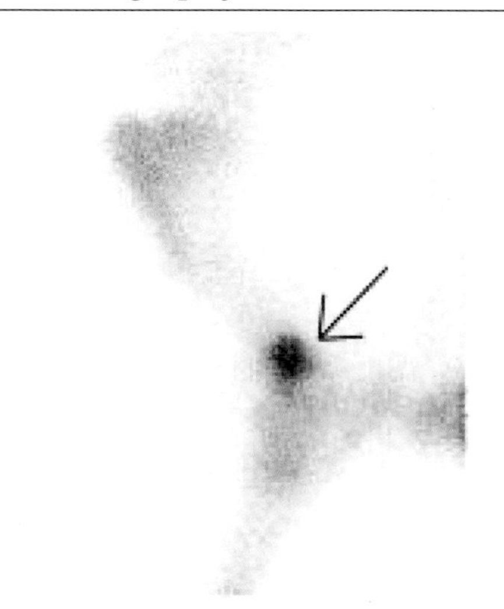

- This scan of an eight-year-old Quarter Horse shows moderate navicular bone uptake on scintigraphy.
- In addition to nuclear scintigraphy as a diagnostic for navicular disease, MRI can help the vet diagnose where that pain is coming from in the majority of horses with pain in the navicular region.
- MRIs can show changes to the navicular bone, such as structural lesions as well as fluid build up, degenerative changes to the coffin joint, and injury or damage to the DDFT or the nearby ligaments.

Nerve Block

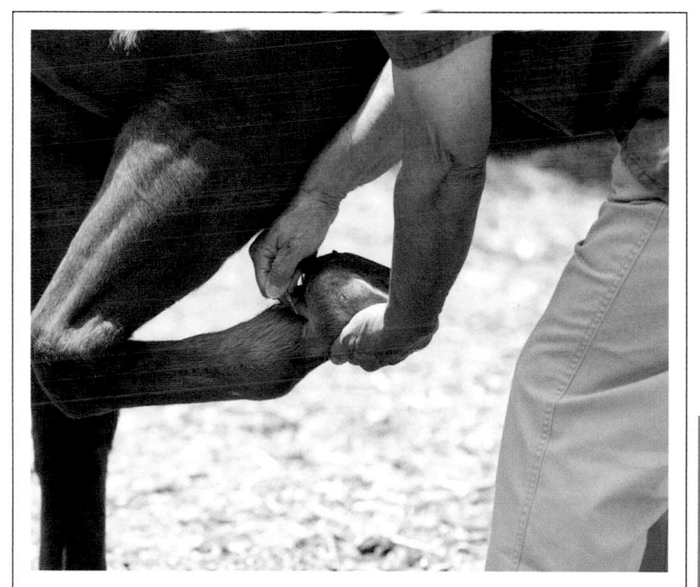

- Nerve blocks are often used to determine if pain is stemming from the heel.
- Soundness after a heel block can be one indicator of navicular disease or syndrome.
- In addition to nerve blocks, if a horse is responsive to navicular bursa analgesia and coffin joint analgesia, these are strong indicators that he may indeed have navicular disease.
- Accurate diagnosis is needed to create a successful management strategy.

NAVICULAR HOOF CARE

Careful trimming and shoeing is part of the management plan for horses with navicular disease

The first goal when shoeing or trimming a horse with navicular disease is to return the hooves to proper balance if they are not already balanced. For example, if the horse has long toe/low or under run heel (see page 10), mismatched feet, or any other hoof balance issues, the farrier will need to address these. Correcting issues and returning the hooves to proper balance can take time, as changes and corrections need to be made slowly. The hoof-pastern axis (see page 8) should be as straight as possible taking into account the horse's conformation. Maintaining healthy heels is also important.

A skilled farrier is imperative. The veterinarian may be able to recommend a farrier experienced in treating horses with

Egg Bar

- Egg bar shoes provide extra heel support and are often used on horses suffering from navicular disease.
- Heart bar shoes, which follow the shape of the frog rather than forming a circle like the egg bar, are another bar shoe option.
- However, steel bar shoes can be heavy, so aluminum versions are a better alternative for horses with navicular disease.

Pads and Sole Support

- The horse shown above is wearing pads.
- Sole support materials, including urethane and dental impression materials, can help spread weight over a greater area, which may benefit some horses suffering from navicular disease or syndrome.
- Like horses being treated for laminitis, horses with navicular pain are sometimes treated with foam pads taped to the horse's hoof.

navicular disease. It's important that the farrier, vet, and owner work together to best manage the condition. In many cases it's helpful for the farrier to view the horse's x-rays or other diagnostic images.

Some vets, farriers, and owners may choose to pull the horse's shoes while the hooves are returning to balance. For managing horses with navicular disease, there are also a number of therapeutic shoeing options, and it often takes trial and error to come up with what works for a particular horse. It's important to allow a couple of months to see if a particular style of shoeing or trimming works for the horse.

Horses with navicular may benefit from a shortened toe for improved breakover, and many shoeing strategies are also aimed at providing heel support. Aluminum egg bar shoes are a common option since they're light and provide some added heel support.

Sometimes a type of shoe or shoeing works for a while but later loses its effectiveness. This is not surprising considering navicular disease is degenerative, so the horse's comfort level and needs will change over time.

So Many Shoes!

- When it comes to shoeing horses with navicular disease or syndrome, experimentation is often part of the process, and a shoe that works for a few months may not work forever.
- Shoes that may benefit horses with navicular disease or syndrome include Natural Balance shoes, wedge pads or shoes, or a wider shoe to help prevent sinking in soft ground.
- Barefoot practitioners have also had success managing horses with navicular syndrome and disease.
- Work with your farrier to keep your horse as sound as possible.

Barefoot

- Although therapeutic shoes are commonly used for horses with navicular disease, some horses can be maintained barefoot.
- The owner, farrier, and vet should work together to decide what's best for the horse.
- Some owners and farriers may choose to pull the horses shoes and give the feet "a break," returning them to balance before applying therapeutic shoes.
- With or without shoes, the goal is to keep the horse comfortable and maintain well-balanced hooves.

MEDICATIONS AND SURGERY

There are medications and surgical options for keeping a horse with navicular disease comfortable

If shoeing and lifestyle modifications are not enough to keep a horse with navicular disease comfortable, then it's time to consider medications. Many horses with navicular disease pain benefit from nonsteroidal anti-inflammatory (NSAID) medications, which also help to relieve pain. There are several choices including the new COX-2 inhibitor firocoxib (EQUIOXX), carprofen (Rimadyl), flunixin meglumine (Banamine), and others, but bute (phenylbutazone) is currently the most commonly prescribed.

There are side effects to NSAIDs like bute, including stomach ulcers, so it's important to only give the horse what's needed to keep him comfortable. Owners can work with

Bute

- Phenylbutazone, or bute, is the most common oral medication for dealing with pain and inflammation in horses with navicular disease.
- Bute is available in several oral forms including tablets, powders, and paste.
- The tablets are usually dissolved or crushed and mixed with feed, although some horses object to the taste.
- Horses that are finicky may be more apt to eat powdered bute, which comes in flavors like citrus.

Injections

- Injections are another way to help manage pain and inflammation in a horse with navicular disease.
- Horses with synovitis (inflammation of the synovial membrane) in the coffin joint may benefit from hyaluronic acid (HA) injections or HA with corticosteroids.
- Corticosteriods can also be injected into the navicular bursa to provide temporary relief.
- Polysulfated glycosaminoglycan (PSGAG like Adequan) given intramuscularly may benefit certain horses.

their veterinarian to find the medication and dosage that's best for the particular horse.

Isoxsuprine is a vasodilator that sometimes is given to improve blood flow and may have some anti-inflammatory effects. Although its effectiveness hasn't been proven, some horses do seem to benefit from it, so the vet may recommend a trial dosage. Injections are another option.

As a last resort, there are surgical options that benefit certain horses. To identify neurectomy candidates, the vet can see if the horse is responsive to analgesia of the palmar digital nerves. If so, he may be a candidate for neurectomy. In addition to traditional neurectomy (discussed below), there is chemical neurectomy, which involves injecting substances over the palmar digital nerves to create a loss of sensation, and there is cryoneurectomy, which freezes a section of the palmar digital nerves to produce a similar effect. Both procedures can lessen pain and lameness for up to three months. Some horses can benefit from navicular suspensory desmotomy, which transects the ligaments to lessen lameness in select candidates.

Palmar Digital Nerves

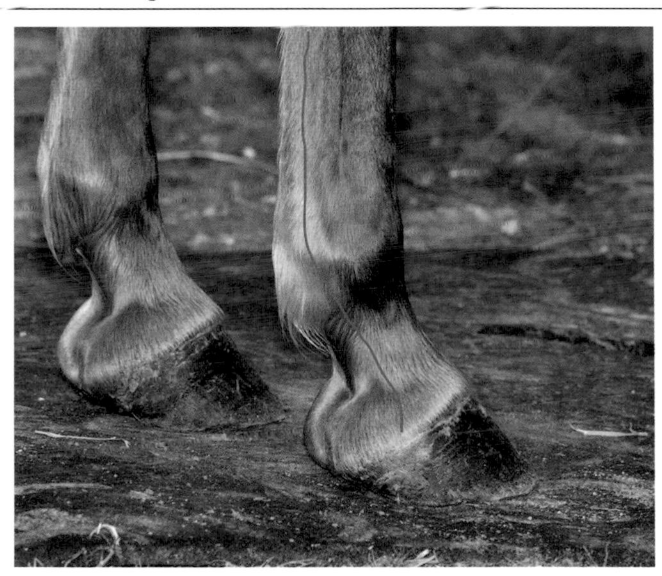

- Palmar means the backside of a front leg, and the palmar digital nerves are the nerves to the back of the foot.
- If a horse is sound after analgesia of the palmar digital nerves, he may be a candidate for neurectomy surgery.
- Neurectomy involves transecting (severing) the palmar digital nerves while leaving the nerves to the toe intact.
- Neurectomy can often provide about a year of reduction in pain and lameness, but the nerves can grow back.

Neurectomy

- Neurectomy can provide relief for some horses with navicular disease, but a horse's living conditions and footing must be considered since he won't have full sensation in his hoof.
- Occasionally the procedure doesn't improve soundness, and side effects are possible including neuromas (where the severed ends form an irritating or painful ball).
- Sometimes injuries or abscesses to the back of the hoof are not recognized after neurectomy due to the horse's lack of sensation in this area.

EXERCISE AND MAINTENANCE

Proper exercise and maintenance can help keep horses with navicular disease comfortable

Exercise and footing play an important role in maintaining a horse with navicular disease. A diagnosis of navicular disease doesn't necessarily mean your horse must immediately be retired. Like horses with degenerative joint disease (DJD), most horses with navicular disease benefit from light, regular exercise. However, horses in high-performance careers, such as barrel racing, dressage, jumping, or roping, may need to have their workload adjusted or may need to switch careers to something with less speed or concussion. Owners should work with their vet to determine how much exercise and what type is advisable for the particular horse and his level of soundness.

Exercise

- Work on hard ground will increase concussion, so try to exercise horses with navicular disease in footing that provides adequate cushion.
- Depending on the horse's soundness level, light exercise, a half hour to an hour, five to seven days a week is often beneficial.
- Encouraging the horse to work off his hind end, as is the goal in collection, can help him carry himself better and place less strain on the front legs.

Footing

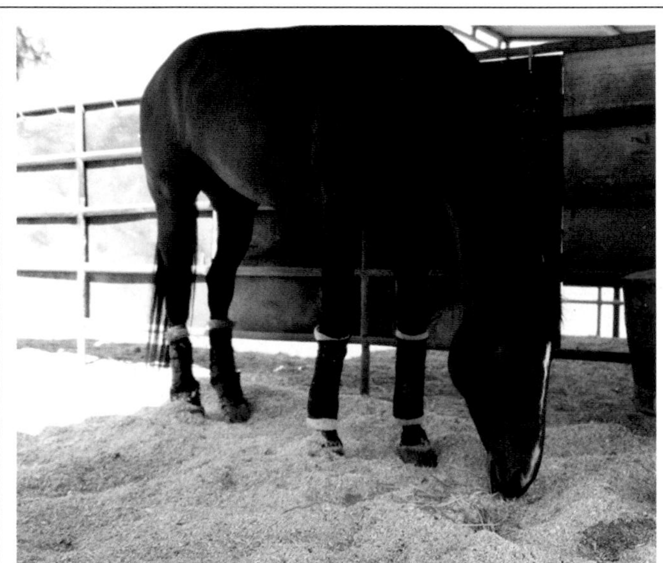

- For horses with navicular disease that spend part of their time in stalls, adequate bedding is important.
- In addition to traditional bedding materials like straw and shavings, there are also cushioned stall mattresses designed to replace bedding and provide the horse extra comfort and support. These provide much more cushion than traditional stall mats, which require additional bedding.
- Providing soft bedding gives the horse added cushion for his feet and encourages him to lie down when he wants to.

If the horse works in an arena, the footing should provide adequate cushion without being too deep, which can strain the soft tissues. Two to three inches of footing is usually considered ideal. For trail horses, rocky or steep footing may need to be avoided if it becomes difficult for the horse.

There will come a time when the horse with navicular disease must eventually be retired from riding; however, that doesn't mean he should stop moving. Free access to a pasture or paddock, or daily turnout is advisable for virtually all horses with navicular disease. A horse with pain issues may be reluctant to move, but movement is critical, and standing around will only worsen existing issues or create new problems.

Exercise improves circulation to the hooves, keeps the horse from getting stiff, and keeps his muscles from becoming too weak from disuse. The muscles support other body structures, and if they're weak from disuse, they will quickly tire or fail, which can place too much strain on joints, tendons, and ligaments. Where turnout is not available, time spent on a hot walker can ensure the horse doesn't stand around all day.

Turnout

- Horses with navicular disease that need to be retired do best in a paddock or pasture, where they're most likely to move around, which encourages blood flow to the feet among other benefits.
- Horses with navicular disease that are still being ridden can also benefit from daily turnout or life in a paddock or pasture.
- However, turnout may not be advisable for horses that have had neurectomies since they don't have full sensation in their feet.

Obesity

- Horses carry most of their weight in their front legs, and front hooves are the ones most frequently affected by navicular disease.
- Overweight horses place even more weight on their front legs.
- Signs that a horse has gained too much weight include fat deposits by the withers, along the neck, behind the shoulders, and above the tailhead.
- In addition to making sure an overweight horse gets proper exercise, owners should consult their vet regarding diet modifications.

BRUISING

The hoof and sole can bruise, causing pain and lameness

It may be difficult to believe that the hoof is capable of bruising, but just on the other side of the hoof wall, sole, and frog is sensitive corium. A bruise is a pocket of blood (hematoma) that forms when blood vessels rupture from trauma to the corium. This can trigger an inflammatory reaction and pain, depending on where the bruise occurs and how severe it is.

Bruises to the hoof wall can occur when the hoof wall hits something hard or is hit by another hoof, and horses with extra thin walls may be especially susceptible. Thin-soled or flat-footed horses (horses with soles that are flatter than most horses) are more prone to sole bruises. Sole bruises are often associated with hard work on hard or uneven terrain. Poor shoeing or trimming can also cause bruises.

Horses with bruises may become suddenly lame, have in-

Hard Ground

- Thin-soled or flat-footed horses worked on rocky terrain are prone to sole bruises. It's possible that even one sharp rock or other object can cause a bruise.
- Hard, frozen, uneven winter terrain can also cause bruising.
- If a horse is prone to bruising, then his working conditions should be carefully evaluated and care taken to work him on more forgiving surfaces.
- However, bruising can also occur without a clear cause.

Foreign Objects

- A rock can lodge along the sides of the frog or in the frog sulcus and cause a bruise.
- It's important to check the horse's hooves frequently, preferably every day, to remove foreign objects and check for any signs of trouble, such as bruising or abscesses.
- Checking hooves daily also allows you to feel if the hooves are hotter than usual or if the horse reacts differently to your touch.

creased heat in the hoof and a bounding digital pulse. Bruising can affect more than one hoof. As with any sudden lameness, the vet should be called to provide a proper diagnosis and treatment plan. Bruises can turn into either bacterial or sterile abscesses (which is a collection of cells and debris without bacteria), so the vet may recommend precautionary measures (abscesses are discussed in the section following this one).

The vet will likely pare away superficial layers of hoof or sole horn to look for a bruise, which may appear as a purple or red area within the corium. If the horse is due for a tetanus booster, the vet will administer one and may also prescribe bute (phenylbutazone) for pain and inflammation. The horse should rest while the bruise heals, and flat-footed or thin-soled horses may benefit from properly applied protective pads and shoes that have the inner edge beveled to reduce pressure on the sole.

Shoeing

- Shoes that are left on too long or that are poorly applied can cause bruising.
- Poorly applied hoof pads can also add pressure to the sole and cause bruising.
- Heel bruises created by shoes are called "corns."
- Corns can form when shoes become loose, are too small, or are too short in the heel area.

Trimming

- Excessive trimming of the hoof wall and sole can cause bruising, so it's important to work with an experienced farrier.
- The vet will take the horse's recent history, including riding activities and any trimming or shoeing, into account when diagnosing a bruise.
- In addition to the causes already discussed, horses with laminitis are also prone to sole bruises.

ABSCESSES

Hoof abscesses can cause the horse severe pain, but prompt treatment usually produces a good outcome

An abscess is an infection within the hoof. Many times no one knows there's an infection brewing until it's progressed to the point that it affects the sensitive laminae and tissues, bringing on sudden lameness and pain.

Abscesses can form when an area of the horse's sole becomes compromised, allowing bacteria to enter, or when the bars of the hoof fold over debris as they grow, trapping bacteria. Puncture wounds can introduce bacteria and then close up before anyone notices (see page 112 for more on puncture wounds).

Horses with poor horn quality or laminitis are especially prone, but an abscess can affect any horse. Once inside the

Draining Tracts and Gravel

- It was suspected that this horse had an abscess before a tract could be found.
- After several days of soaking the hoof in warm water and Epson salts, the abscess broke through near the coronet and began draining.
- It's a myth that tiny pieces of gravel enter at the bottom of the hoof and travel up causing abscesses like this.
- This infection traveled up and opened at the top of the hoof rather than along the sole, which can also happen.

Epsom Salts

- Soaking the hooves in warm water and Epsom salts helps draw out the infection.
- Epsom salt soaks may be recommended to draw the infection out before a tract is created or to help an existing tract drain completely.
- Hooves can be soaked using a common bucket, but if the horse moves, the bucket may be tipped over.
- There are also commercial soaking boots that make the process easier.

moist, warm hoof, the bacteria multiply and produce tissue-damaging toxins. The horse's inflammatory response to try to kill the infection can also damage tissues. Soon, a large pocket of pus forms; pressure may create a draining tract for the pus at the sole or near the top of the hoof.

Signs of abscess include a bounding digital pulse, a swollen lower leg, sensitivity to hoof testers, heat in the hoof, and lameness possibly so severe the horse won't walk or bear any weight on the affected foot.

If pus is already draining from the hoof, or the vet suspects an abscess, the goal is to create a good draining tract with a hoof knife and clean the pus out with antiseptic solution. Sometimes the pus will pour out, and other times the hoof will need to be soaked with warm water and Epsom salts to draw it out.

After draining, the hoof should be bandaged while it heals unless the tract is high up near the coronet. A pain killer and anti-inflammatory such as bute (phenylbutazone) will likely be prescribed, and possibly a tetanus booster and antibiotics. In severe cases, the infection can affect the bone.

Hoof Bandaging

- As the hoof is healing, it may need to be bandaged.
- Bandages can be created using wrapping materials and duct tape, a commercial hoof boot, or a combination of the two, or a hospital plate can be added to a horse shoe.
- The draining tract is usually packed with cotton soaked in antiseptic solution before the hoof is bandaged.
- Horses recovering from abscesses should be put on stall rest in a clean stall or paddock.

Living Conditions

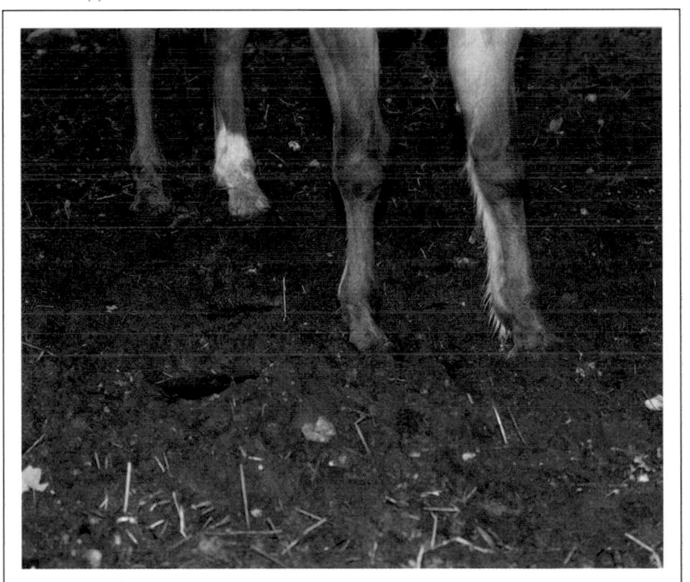

- A horse living in unsanitary conditions, such as a wet pen with manure and urine, will have a softer sole, making him more prone to bruises and abscesses.
- Hooves that aren't regularly cleaned will also hold more dirt, debris, and bacteria.
- Proper maintenance and regular trimming and shoeing can help prevent abscesses from occurring.
- Horses promptly and correctly treated for an abscess usually make a full recovery.

WHITE LINE DISEASE

White line disease, or hoof wall separation, is extremely common and must be caught and treated early

White line disease (also called "seedy toe" or hoof wall disease) is a common problem that can have devastating effects. The name white line disease is a little misleading because the problem is actually in the nonpigmented inner part of the hoof wall, just outside the white line.

Although white line disease occurs throughout the country, horses living in humid wet areas seem especially susceptible. Hooves weakened by moisture or poor nutrition may receive extra stress from unbalanced feet, such as horses with long toe/low or under run heel. The weakened wall starts to separate from the sole, and bacteria and fungi can enter and cause further deterioration. Horses with existing problems

White Line Disease

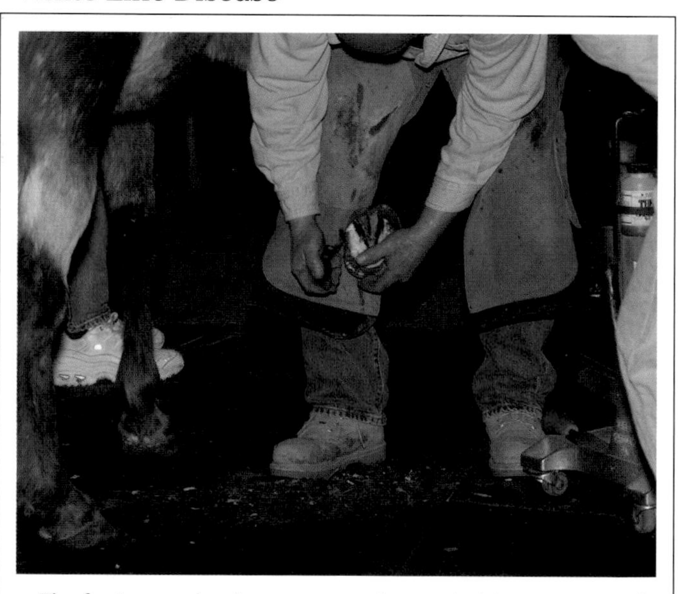

- The farrier may be the first to spot the beginning stages of white line disease.
- Poor shoeing, trimming, or unbalanced feet can make a horse more susceptible to white line disease.
- Any underlying causes such as unbalanced feet, abscesses, or laminitis should be treated along with the white line disease.
- The vet and farrier will determine how much exercise the horse can handle during recovery.

Debridement

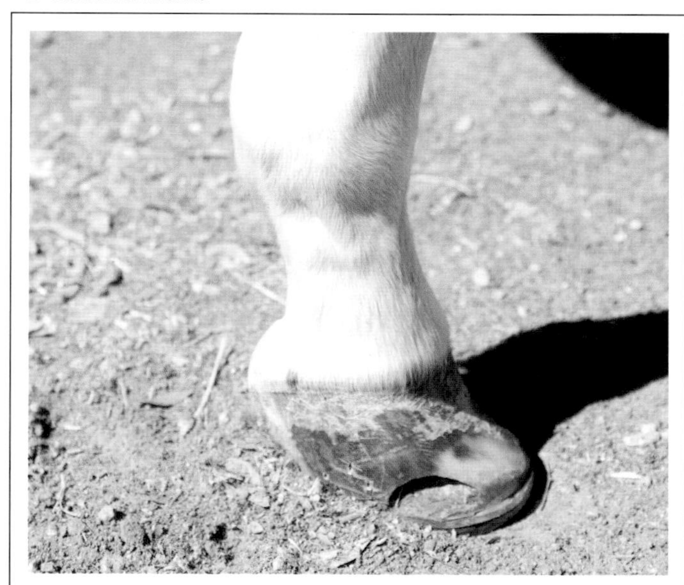

- In horses with white line disease, the damaged wall will be removed (shown above).
- Shoeing will be based on individual needs, and supportive egg bar or heart bar shoes are often used during healing.
- If the hoof wall can't support nails during healing, glue-on shoes are also an option.
- Using acrylic products to fill in the missing hoof wall is not advisable, as it can trap in bacteria.

such as abscesses, bruises, laminitis, and hoof cracks are also more susceptible to white line disease.

White line disease can strike virtually any horse and occur in one foot or more. The wall and sole will begin to separate somewhere in the nonpigmented area along the toes or quarters, and grayish white chalky material may be visible where the wall and sole meet. Other signs may include heat in the hoof, sensitivity to hoof testers, change in hoof growth, and change in hoof shape, such as a hoof that becomes flatter or has bulges and/or concavity. The farrier may be the first to discover white line disease as he trims the hoof.

Lameness will not appear until the sensitive laminae are affected. In severe cases, there can be rotation of the coffin bone, as happens in cases of founder (see page 140). X-rays can check for rotation and often will show separation. Treatment involves debridement, where the damaged hoof wall is removed until healthy hoof wall is exposed. Continued debridement and supportive shoeing usually produce a good outcome in cases where rotation hasn't taken place, but it can take a year for the hoof to completely grow out.

Environment

- Environmental factors such as changes in moisture levels or constantly wet settings can weaken hooves.
- Avoiding extreme fluctuations in moisture, keeping pens clean, and making sure horses' hooves aren't constantly wet will help prevent white line disease.
- Catching white line disease early means a quicker recovery and better prognosis.
- For a horse recovering from white line disease, a dry and clean environment is critical.

Prevention and Communication

- In order to prevent problems, communicate with your farrier about the health of your horse's hooves.
- Things like poor nutrition can create weak hooves, making a horse more susceptible to white line disease and other problems.
- If a horse has weak hooves, the owner should work with the vet to create an optimal diet based on the horse's needs.
- Hoof supplements may be recommended for horses with weak hooves or white line disease.

KERATOMA, CANKER, AND QUITTOR

While these hoof conditions are uncommon, it's important to recognize them and provide veterinary treatment

Keratoma, canker, and quittor are relatively uncommon hoof conditions but still do occur. If they are recognized early, treatment is usually successful.

A keratoma is a mass of keratinous cells within the inner hoof wall. Hooves are made from keratin, but this abnormal growth creates a tumor that can grow large enough to cause pressure within the hoof, resulting in pain and lameness. Eventually, it may grow large enough to displace the coffin bone.

Symptoms are often similar to a coffin bone fracture or abscess. X-rays are used for diagnosis, and the section of hoof wall with the keratoma will need to be removed. Supportive shoeing, possibly including a hospital plate, is used while

Keratoma

- A keratoma is a keratinous tumor that can be round or cylindrical.
- Keratomas can occur anywhere in the toe or quarters of the hoof wall from the coronet down.
- What causes keratomas isn't completely clear, but it may occur as the result of a sole abscess, or from trauma or injury to the hoof.

Canker

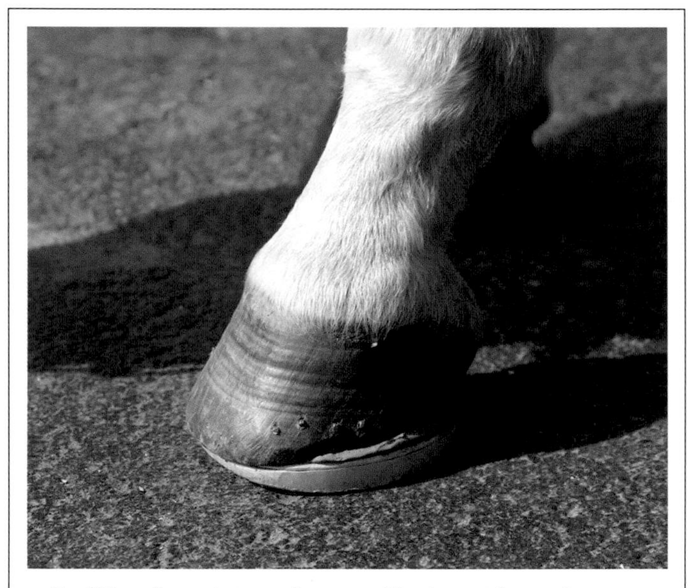

- Traditionally, wet, unsanitary environments and poor hoof care were blamed for causing canker, but it may be caused by trauma that allows bacteria into the horn.
- Treatment for canker can take anywhere from ten days to several months in persistent cases.

the wall grows out. Prognosis is usually good, and keratomas don't often reoccur.

Canker is a bacterial infection of the sole or frog that creates abnormal, excessive growth. Canker is usually seen as a white or yellow, stinking, oozing, spongelike mound in the sulcus area of the frog.

To treat canker the vet will usually debride the area (remove affected tissue) and prescribe a topical antibiotic/antimicrobial. The area will likely need repeated debridement and a dry bandage until it heals.

Quittor, sometimes mistaken for an abscess because it usually creates draining tracts near the coronet, is an infection of the lateral cartilages usually caused by a wound.

The area may be warm, swollen, and painful, and lameness varies. X-rays can often confirm diagnosis. Topical and oral antibiotics along with a tetanus booster shot (if needed) may be the first round of attack, but because the lateral cartilages don't have much blood supply, the medication may not work.

Quittor

- Quittor can be caused by a puncture wound (such as a nail), a laceration (for example, a wire cut near the coronet), or repeated interference injuries to the coronet area.
- Often the veterinarian must surgically remove the infected parts of the lateral cartilage. Postsurgical care will likely include rest, and wrapping and packing the affected area with gauze and topical solution.
- Healing can take considerable time, and prognosis often depends on whether the hoof is permanently deformed.

Draft Breeds

- Canker and quittor are more common in draft breeds.
- Canker mostly occurs in the back hooves, while quittor is more common in the front feet.
- Quittor used to sometimes be called "treads" because draft teams often wore studs and a horse might tread on the hoof of the horse on his side, causing a coronet injury, which could lead to quittor.

PEDAL OSTEITIS

Pedal osteitis describes changes to the solar margin of the coffin bone but can have many causes

Pedal osteitis is one of those tricky terms like *navicular syndrome*. It literally means coffin bone (pedal) inflammation (osteitis), but it's actually a descriptive term meaning that demineralization (thinning) of the solar margins of the coffin bone can be seen on x-ray. It's not a specific disease, and the changes seen on x-ray could have occurred years ago. The cause may be known or unknown, and horses with these x-ray findings may or may not be presently lame.

Widening of the vascular channels and demineralization of the coffin bone makes it lose its smooth contour, which shows up on x-ray. Anything that causes excess blood or pressure to the area can lead to these bone changes. Pedal

Concussion and Bruising

- Work on hard ground—especially repeated or fast work—should be avoided.
- Any type of trauma to the feet that leads to increased blood and pressure in the hoof could potentially lead to demineralization (thinning) of the solar margin of the coffin bone, as is seen in pedal osteitis.
- Chronic bruising, as can happen in horses with thin soles, flat feet, or long toes, may also lead to the bone changes seen via x-ray in horses with pedal osteitis.

Hoof Testers

- Hoof testers are used to help the vet determine if lameness comes from the foot and if so, where in the foot.
- Hoof testers are part of the diagnosis for almost all hoof problems.
- However, it's important to remember that pedal osteitis isn't a diagnosis in and of itself, rather just an indication of bone loss to the coffin bone.
- If possible, the primary cause of pedal osteitis should be determined for accurate treatment.

osteitis may be associated with laminitis, coffin bone fracture, infections (from an abscess or puncture wound), keratomas, chronic bruising, or other traumas.

Pedal osteitis may be discovered during a prepurchase exam or because a horse shows lameness, which can be chronic and intermittent similar to navicular disease or a fracture to the coffin bone. Hoof testers and nerve blocks may isolate the pain to the hoof, and pedal osteitis may then show up via x-ray.

Treatment should be aimed at the primary cause of pedal osteitis, assuming it's still active. For example, if there's an infection or the horse suffers from laminitis, these problems should be treated. If he is susceptible to chronic bruising, then shoeing and lifestyle changes will need to be made.

If the primary cause is no longer active or can't be determined, then a nonsteroidal anti-inflammatory like bute (phenylbutazone) may be given as a pain reliever. In some cases, neurectomy of the palmar digital nerves may provide the horse with at least temporary relief of chronic pain (for more information on neurectomy, see page 161).

Pedal Osteitis

- If the vet says the horse has pedal osteitis, it means that demineralization of the solar margin (bottom part) of the coffin bone can be seen via x-ray.
- Horses with these x-ray findings may or may not be lame, and the changes may have occurred years ago, or the primary cause may still be active.
- Some experts believe the term pedal osteitis shouldn't be used because it implies it's a disease with known causes and treatments.

Shoeing

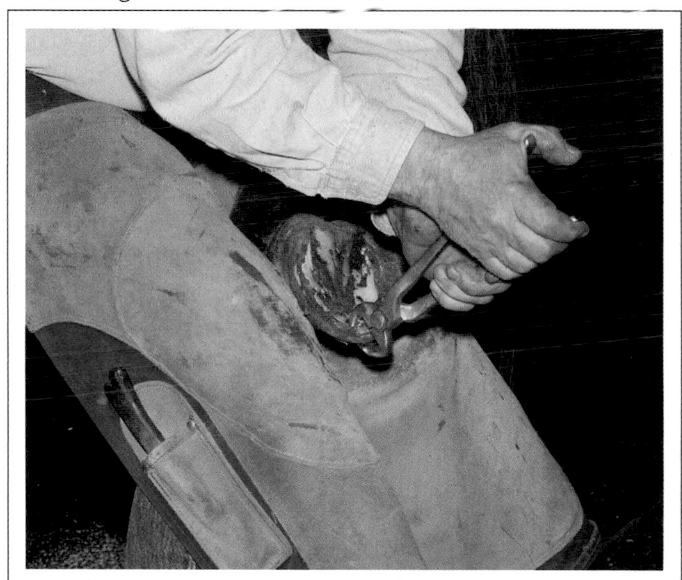

- Shoeing a horse with pedal osteitis depends on the primary cause.
- If a horse is susceptible to chronic bruising, then trimming and shoeing should be aimed at correcting any unbalance and protecting the soles of thin-soled or flat-footed horses.
- The best way to prevent pedal osteitis and other permanent damage to the hoof and its structures is to practice good hoof care with regular trimming and shoeing, and to promptly provide veterinary treatment for any injuries.

HOOF CRACKS

Hoof cracks need treatment tailored to their specific cause, location, and depth

Hoof cracks generally travel vertically up and down the hoof and can be no big deal or a huge problem, depending on several factors. Hoof cracks are categorized by their location—toe, quarter, heel, or bar; their depth—superficial or deep; their length—complete from the coronary band to the ground, or incomplete (and their origin—whether they start from the ground up or the coronary band down. These factors dictate treatment. Deep cracks can be especially worrisome, as they may become infected and affect the sensitive structures inside the hoof, resulting in severe lameness.

Hoof cracks can be caused by a variety of factors, including too-dry hooves (brittle), too-wet hooves (soft and weak),

Hoof Crack

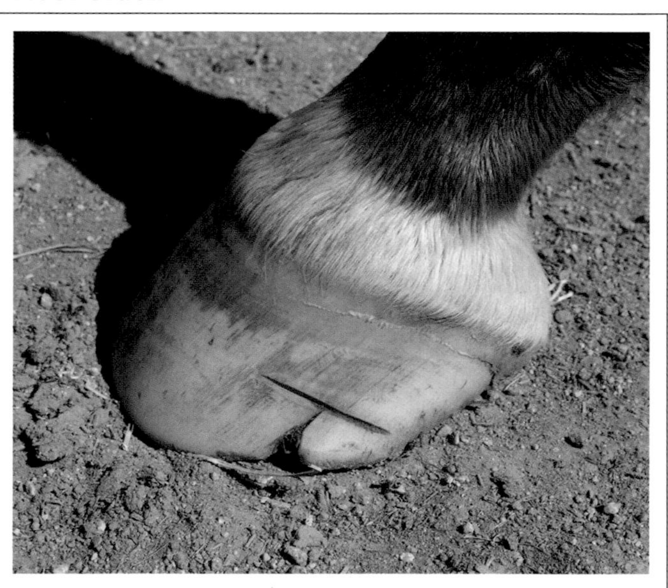

- The horse in the photograph above suffers from a persistent crack along the quarter (side) of his front left hoof.
- Because this hoof toes in due to arthritis in the horse's left knee, it places added weight on the outside of the hoof. In addition, an old hoof injury has made this hoof slightly deformed (neither of these conditions are readily visible in the photo of the crack).
- The farrier has made a horizontal notch to keep the crack from traveling upward.

Crack from Neglect

- This crack is due to neglect.
- This hoof has not been regularly trimmed and has been allowed to grow far too long.
- In addition, the hoof shows some deformity, as seen where the coronet curves up along the sides.
- Unfortunately, hooves like this take time to return to balance. The farrier cannot trim an overgrown hoof back all at once without causing pain and additional damage. Instead, the farrier must slowly trim the hoof back over many months.

unbalanced hooves (including under run heels), wall separation, laminitis, excessive growth, trauma or injury to the coronary band or coronet, infection, work on hard ground, and poor trimming or shoeing.

Treatment goals include addressing the cause, as cracks can become a chronic problem, and stabilizing the current crack. Superficial cracks are often treated by balancing the hoof and then using full bar shoes with clips.

If the crack is infected and affecting sensitive tissues, the damaged wall will need to be removed and the area debrided (all damaged materials removed). Once any infection present has been completely cleared up, serious cracks may be repaired using suture materials, screws, clamps, plates, wires, or various synthetic repair materials, such as acrylic, fiberglass, epoxies, or rubber.

Serious hoof cracks can take many months to grow out. The best way to prevent hoof cracks is to maintain well-balanced feet that receive regular, quality trimming and shoeing. Also provide a healthy diet for your horse and properly attend to any injuries.

Shoeing

- Trimming and shoeing after a crack must address the cause, rebalance the hoof, and stabilize the crack.
- Superficial cracks can often be stabilized using a bar shoe with clips.
- To move weight bearing away from a crack, some farriers use pads.
- Some farriers also burn or rasp a horizontal notch at the top of the crack to keep it from becoming longer.
- Superficial cracks can usually be treated by the farrier, but deeper cracks or cracks leading to infection need veterinary attention.

Repair Materials

- There are many repair materials that can be used on deeper cracks.
- One type of repair technique, fabric lacing, uses acrylic adhesive that will actually grow out with the hoof wall and can be trimmed and shod.
- However, repair techniques like this can be expensive.
- The vet and farrier should work together to decide the best treatment plan for deep cracks.

SELENIUM TOXICOSIS

This disease can cause the hoof wall to separate and, in severe cases, fall off

Selenium is a trace mineral essential to horses and humans in small amounts. However, too much selenium can be toxic and produce devastating effects. Horses that have ingested too much selenium are said to have selenium toxicosis, or alkali disease.

Many feeds and supplements contain selenium, so over supplementing your horse can cause selenium toxicosis. Selenium toxicosis can also result from environmental contamination, such as too much selenium in the water. However, many horses get selenium toxicosis from their pasture. Some areas of the country have higher levels of selenium in the soil. States that may have selenium rich soil include North and

Feed

- It's important to monitor the horse's pasture and provide additional hay when quality pasture grass runs low.
- Horses usually won't eat indicator plants when there are better things to eat.
- If you suspect selenium toxicosis, remove your horse from the pasture until the vet can make an accurate diagnosis.
- If you live in an area that may have selenium rich soil, you can have your soil tested by a local lab or university.

Supplements

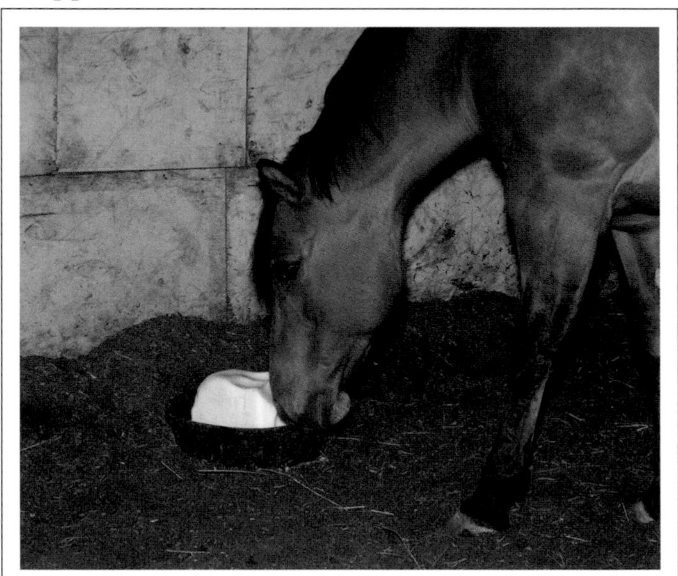

- Some salt licks are fortified with trace minerals including selenium.
- Although uncommon, horses and foals can be selenium deficient.
- Selenium deficiency is most likely in areas with selenium-deficient soil, where the horses eat grass and hay that's homegrown, and receive no outside or commercial feeds.
- As always, your vet is the best person to advise you on dietary considerations for your particular horse and area.

South Dakota, Montana, Wyoming, Colorado, and Utah.

There are certain plants called indicator plants that may grow in selenium rich areas and contain a high amount of selenium. These plants include milk vetch, woody aster, prince's plume, salt bush, and goldenweed. If you notice these plants on your property, or you live in an area that may have selenium rich soil, talk to your vet about selenium toxicosis, pasture safety, and a proper diet for your horse.

The first signs of selenium toxicosis are usually a dull coat and loss of mane and tail hair. Lameness, stiffness in joints, and other signs of discomfort may also be present. Dry, cracked, and brittle hooves are a common symptom of selenium toxicosis. In severe cases hoof abscesses and a horizontal separation of the hoof wall may occur, possibly leading to loss of the entire hoof wall.

The vet may take hair or blood samples to confirm the diagnosis. Pastured horses should be removed from the pasture, and dietary changes will likely be in order. The vet will address abscesses, and hoof boots and other emergency care should be used if the hoof wall has sloughed (fallen) away.

Signs

- Excess selenium can alter cell function.
- Cells that form keratin, such as those involved in the production of hair and hooves are especially sensitive.
- The first obvious sign of selenium toxicosis is usually loss of mane and tail hair as if someone took clippers to your horse. This is usually accompanied by a generally dull coat.
- The next sign is cracked and brittle hooves. In severe cases a separation ridge may develop below the coronet and the entire hoof capsule begins to come loose.

Blood Samples

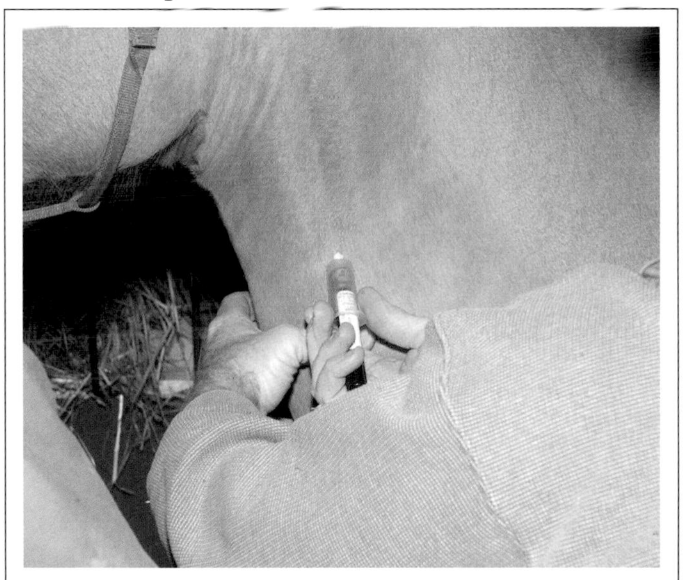

- A blood sample can also help identify a horse or foal that may be selenium deficient.
- More commonly, a blood sample is used to confirm a diagnosis of selenium toxicosis.
- If horses with selenium toxicosis are removed from pastures with seleniferous forage, they may recover without permanent damage.
- How quickly horses recover from selenium toxicosis varies based on the severity.

MUSCLE ISSUES

Fibrotic myopathy, sweeney, and tying up can affect your horse's muscles and cause lameness

Muscles, like other parts of a horse's legs, can suffer from injury, strain, and a variety of conditions that lead to lameness. Here we'll look at three muscle-related conditions.

Fibrotic myopathy can result after injury to the horse's thigh muscles—most often the flexor muscles of the stifle, specifically the semitendinosus muscle. Muscle tearing or chronic strain can result in inflammation and fibrosis—the formation of scar tissue. Horses with fibrotic myopathy have scarring that restricts the muscle. When moving, the horse slaps the affected hind limb down and will also have a short stride with minimal lift. In some cases physical therapy can help stretch the affected muscles, but in severe cases surgery is

Fibrotic Myopathy

- Fibrotic myopathy is fibrosis of the thigh muscles after an injury or trauma, which creates gait dysfunction. Sometimes the resulting scarring and calcium deposits in the muscle can be felt.
- Horses that perform sudden stops and turns may be more susceptible to this type of injury.
- Fibrotic myopathy usually affects the semitendinosus muscle shown in this photo of a healthy horse.
- Treatment may involve surgical transection of the semitendinosus tendon.

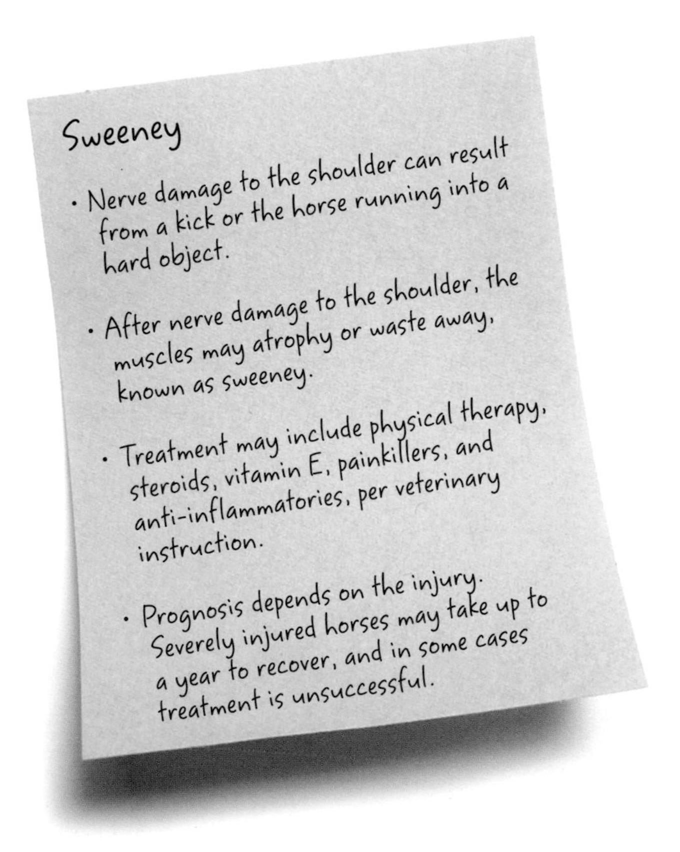

needed to help release the restriction.

Injury to a horse's shoulder may result in nerve damage, leading to muscle atrophy known as sweeney or sweeney shoulder. Sweeney usually is the result of injury to the suprascapular nerve that supplies the large supraspinatus and infraspinatus muscles. These shoulder muscles waste away, shrinking in size, leaving the scapula clearly visible.

Occasional tying up, also called exertional rhabdomyolysis, may occur in horses with any combination of the following: overexertion, dehydration, vitamin E deficiency, or electrolyte imbalance. Exertional rhabdomyolysis means the muscles are breaking down and releasing certain muscle enzymes. These muscle enzymes can be measured for diagnosis and can also cause the urine to become brown tinged. Signs of tying up include tightness in the gluteal muscles, stiffness or reluctance to move, and indications of distress, such as sweating, rolling, and an increased heart and/or respiratory rate. If a horse is tying up, he should be kept still. The vet will administer treatment aimed at controlling pain, rehydrating, and improving electrolyte balance and blood flow.

Tying Up/Exertional Rhabdomyolysis

- Overexertion and dehydration play a role in exertional rhabdomyolysis. Some Thoroughbreds, Standardbreds, and Arabians are prone to recurrent exertional rhabdomyolysis, or tying up.
- Treatment for horses suffering from recurrent exertional rhabdomyolysis includes light work, reducing stress, and a low starch diet.
- A metabolic condition called polysaccharide storage myopathy (PSSM) affects some heavier breeds, predisposing them to recurrent tying up. There are several other genetic causes in addition to PSSM.

Direct Trauma

- A kick from another horse can cause muscle damage and injury to other soft tissues.
- A kick to the shoulder can also cause nerve damage leading to sweeney.
- If your horse has been kicked and shows signs of pain, lameness, or injury, consult your vet.
- Providing horses plenty of room, separating them at feeding time, or feeding them far apart can reduce the incidence of kicking.

STRINGHALT AND SHIVERS

Neuromuscular conditions like stringhalt and shivers can cause hind limb lameness and gait dysfunction

Nerves stimulate muscles, and neuromuscular issues affect both the nerves and the muscles. Here, we'll look at stringhalt and shivers, two neuromuscular conditions that lead to severe hind-end gait abnormalities. While these conditions can be mistaken for upward fixation of the patella (see page 121), there are differences in the movement of the hind limbs.

Stringhalt affects the nerves to the muscles resulting in hyperflexion of one or both hocks. The affected limb(s) suddenly jerks up, either occasionally or with every step. Sometimes the hyperflexion is so severe the hoof or cannon bone hits the horse's belly.

There are two types of stringhalt. Sporadic stringhalt occurs

Stringhalt Treatments

- Stringhalt is usually most obvious when the horse is first asked to move, when he's backing up, or when he's walking, and signs of stringhalt may not be present at the trot or canter.
- Some horses recover from stringhalt with rest alone.
- Other treatments that are sometimes effective include sedatives to reduce anxiety and surgical procedures to cut portions of tendon or muscle in the hock area.

Toxic Plants

- Horses usually will stay clear of the plants implicated in Australian stringhalt unless other forage is unavailable.
- As always, it's important to monitor pasture quality and supplement pastures with quality hay when grass runs low.
- Horses affected by stringhalt should be removed from pasture and the pasture inspected for the plants known to cause stringhalt.
- Photos of plants that are toxic to horses are readily available online, and it's important for horse owners to familiarize themselves.

in individual horses. Exact causes are unknown, but some horses develop stringhalt after injury to the front of the hock. The second type is known as Australian stringhalt and usually affects multiple horses at one location. In these cases certain plants are usually to blame, including types of cat's ear, dandelion, and mallow plants, which are found throughout the United States. Other possible culprits include types of flatweed, vetch, and sweet peas. Compounds in these plants may damage nerves, but the exact process is still unknown. Treatments are discussed in the photos below.

Shivers is a mysterious neuromuscular disease resulting in involuntary muscle spasms in the pelvis, tail, and one or both hind legs. Affected horses will often lift a hind leg, and the leg will be held away from the body as it shakes—the horse unable to control its actions. Signs are often most evident when the horse is asked to pick up his hoof for cleaning or asked to back up. Shivers is a progressive condition, resulting in muscle wasting and weakness. There is no cure, though in the early stages some horses benefit from a high-fat and low-carbohydrate diet.

Shivers Symptoms and Progression

- Horses with shivers will often hold the affected hind leg out, which differentiates it from stringhalt.
- Other signs of shivers include atrophy of the thigh progressing to general muscle atrophy, and hind limb stiffness and weakness.
- Horses may improve with rest, only to have symptoms return with exercise.
- The disease can progress very rapidly or slowly over many years, but there is no cure.

Shivers Causes

- Shivers has a poor prognosis, and the condition worsens over time.
- The disease most often affects draft breeds, but can less commonly affect lighter breeds and is rarely seen in ponies.
- There is possibly a hereditary component to shivers.
- While an exact cause is unknown, shivers could be linked to a neurologic pathology or could be linked to complications from infection or linked to diet.

SKIN CONDITIONS

Although not an emergency, skin conditions like scratches and rain rot are a common problem

Horse legs often end up exposed to moisture and mud, which are frequent culprits in skin conditions. Skin problems can be caused by bacteria, fungi, mange, cancer, or reactions to chemicals, plants, sunlight, or insects. Often the owner will notice scabs, loss of hair, or crud on the lower legs.

It's usually safe to try treating skin conditions using an antibacterial/antifungal shampoo or spray, but if the condition worsens, spreads to other animals, or doesn't respond within a week, contact the veterinarian. Certain conditions can mimic others. The veterinarian can pinpoint the true cause by examining the leg and taking a skin scraping, fungal culture, or skin biopsy.

Scratches

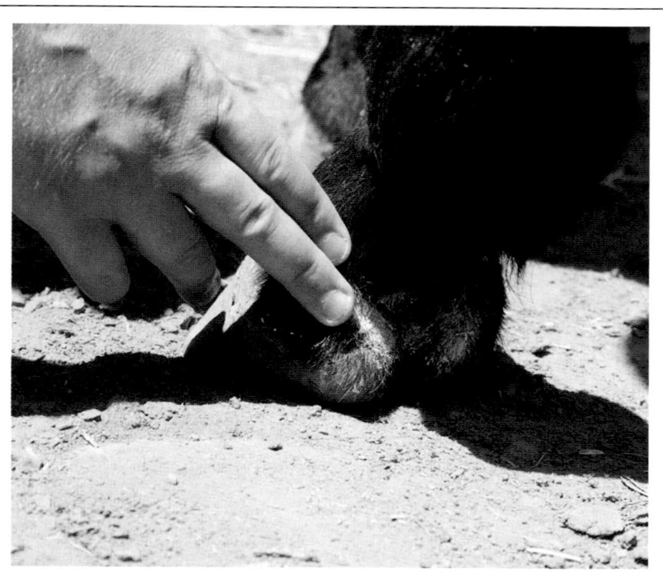

- Horses with scratches should be removed from wet, unsanitary conditions.
- Wash the affected legs with shampoo, dry them and then gently clip the hair around the irritation.
- Although certain home remedies should be avoided, Desitin is sometimes effective on scratches, as are zinc oxide or ichthammol ointments.
- After a few days of treatment, the scabs or crusts should fall away and the area begin to heal; otherwise, consult the vet.

Cannon Crud

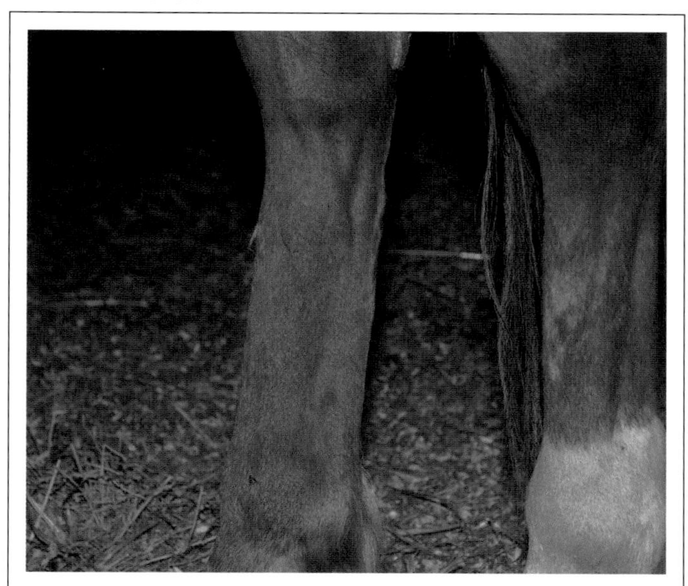

- Cannon crud, cannon keratosis, or scurf, often appears on the cannon bones of hind legs.
- Signs may be mild and look like dandruff, or the leg may have clumps of gunk and hair that can be scraped off.
- If the condition is not caused by fungi or bacteria, the vet may prescribe topical corticosteroids, vitamin A cream, or a defoliating cream or solution.
- A drying shampoo may also be recommended.

Scratches—also called pastern dermatitis, greasy heel, mud fever, or cracked heels—appear as scabs, crusts, or cracked skin on both hind legs just above the heel. Generally, the area becomes irritated during periods of wet and then dry weather, cracking the skin, which can make it susceptible to bacteria or other irritants.

Rain rot or rain scald is a common skin infection that can affect the horse's legs and other parts of his body, including his back. The organisms that cause rain rot are present in warm, damp environments and take advantage of damaged or irritated skin. Rain rot appears as scabs matting the lower layer of hair. When the scabs are removed, the skin is irritated, and there may be pus present.

Rain rot can spread from horse to horse, or be spread by insects or shared equipment. Once a horse or barn is affected, rain rot may reoccur when conditions are right. Unsanitary conditions and damp stalls with poor ventilation create the perfect environment for rain rot.

Specific treatments for both rain rot and scratches are discussed in the photos below.

Rain Rot

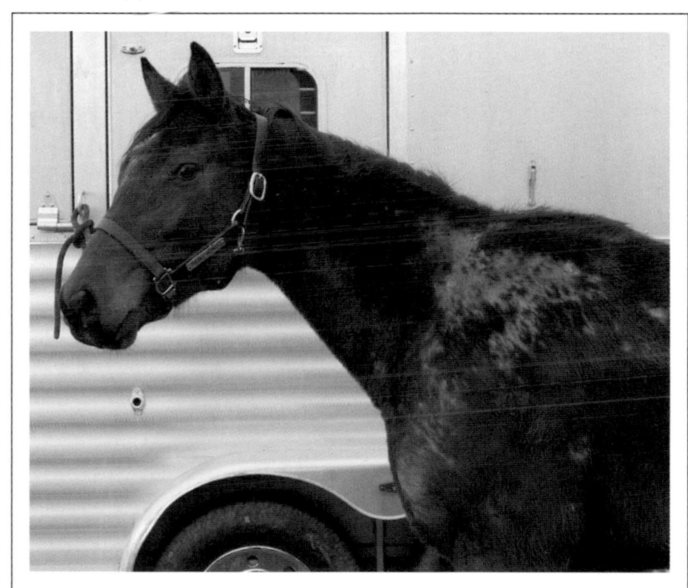

- Horses with rain rot should be isolated from other horses and moved to a dry environment.
- The horse should be washed with an antimicrobial/antibacterial shampoo, and the scabs gently removed.
- Antimicrobial/antibacterial shampoo washes are recommended daily for the first week and then twice a week until the rain rot clears up.
- Systemic antibiotics may be required.

Shampoos

- Although wet conditions are often to blame for skin problems, washing the affected areas with antibacterial/antifungal shampoos is usually helpful; just be sure to rinse the horse well and dry him completely.
- Look for antibacterial/antifungal shampoos and sprays with ingredients like chlorhexidine or iodine.
- If the horse has a thick winter coat or only the legs are affected, it's best to spot wash rather than wet the entire horse.

PREVENTING SKIN CONDITIONS

Taking preventative measures can reduce the scars, pain, and irritation skin problems cause

Damage and hair loss resulting from skin problems can result in white hair growing in where colored hair was before, leaving a permanent blemish. Skin conditions are also irritating or painful to the horse and can be difficult to clear up. For these reasons, prevention should always be the goal.

Providing a sanitary, comfortable place to live is tops in preventing skin conditions. Urine, manure, and mud not only house bacteria and fungi, they also attract flies and other insects. Keeping the horse healthy overall with a proper diet, exercise, and regular veterinary care is also important and helps keep his immune system strong to fight problems and infections.

Unsanitary Conditions

- Muddy or dirty pens are often to blame for skin conditions.
- Always provide horses with shelter from rain and snow.
- Manage footing and drainage so that horses have dry areas to stand and corrals are not overtaken by urine or mud.
- Clean stalls and pens daily to remove manure and soiled bedding, and dispose of waste away from the horse's pen.

Equipment

- Skin conditions like rain rot can be spread by shared equipment.
- If possible, each horse should have his own set of brushes and his own leg wraps, saddle pads, and tack.
- Brushes and equipment should be disinfected following an outbreak.
- Equipment should also be cleaned regularly.

It's important to groom the horse regularly with clean brushes. During your grooming sessions, run your hands over the horse's coat, feeling with your fingers for scabs, clumps, or other signs of problems. Treating skin conditions early helps them clear up faster and avoids scarring. And while occasional baths can be a good thing, over washing the horse may irritate his skin or dry it out.

Some pink skinned horses are susceptible to sunburn. They can benefit from added shelter, sunscreen, and protective wear, such as fly masks, day sheets, or mesh leg protection.

ZOOM

Many horses have insect sensitivities that can cause severe itching and hair loss to the lower legs. Protecting a horse from insects takes diligence on the owner's part and can include topical repellents, barrier methods, keeping the horse indoors during peak insect activity, and providing fans or other ventilation inside the barn.

Insect Sensitivity

- Insect sensitivity can drive a horse crazy with itching and irritation, and in severe cases veterinary treatment will be needed.
- Determining which insects a horse is sensitive to can help avoid contact. For example, a horse can be kept indoors when the insect is most active, such as dusk and dawn.
- Fly sprays and rub-on repellents should generally be applied until the leg is damp.
- Follow repellent directions, and reapply frequently per manufacturer instructions.

Protection

- In addition to topical repellents, barrier methods help prevent the insects from coming into contact with the horse.
- Mesh fly boots can keep flies and other insects from biting a horse's legs.
- However, mesh fly boots often slip down and bunch up, so it's important to monitor and readjust as needed.
- There are also spot-on treatments that last several weeks but are often best used in addition to other methods.

DEVELOPMENTAL DISORDERS

This family of conditions commonly impacts the limbs of growing foals

Developmental orthopedic disorders (DOD) can affect growing horses' conformation, movement, and soundness. DODs often involve the growth plates. Cartilage becomes new bone (ossifies) at the growth plates (at the ends of bones in growing horses).

DODs are acquired rather than congenital (present at birth) and include angular deformities, which occur in the growth plate and affect the angle of the limb; flexural deformities (i.e., contracted tendons); physitis/epiphysitis, which affects the long bones and involves swelling around the growth plate; osteochondrosis (OC), which refers to cartilage that doesn't properly ossify as it should in a growing horse and

Flexural Deformity: Contracted Tendons

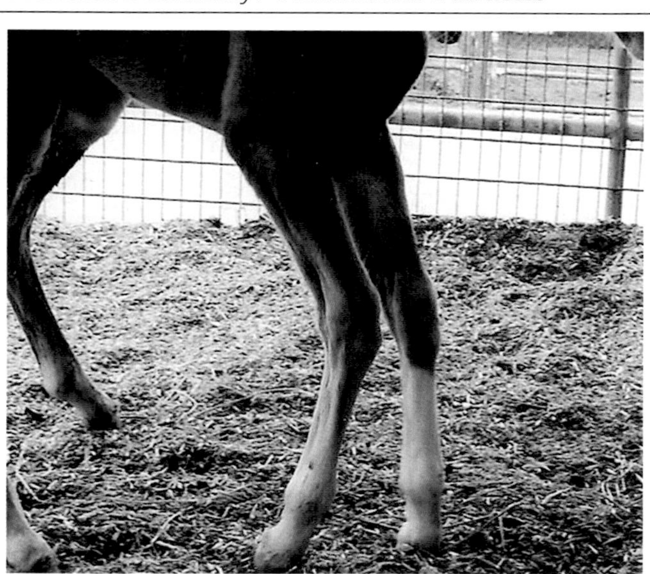

- A foal can be born with a congenital flexural deformity or have an acquired flexural deformity.
- There are many possible causes for congenital and acquired flexural deformities.
- "Contracted tendons" is the common layman's term for flexural deformities in the legs, such as seen in the photograph above.
- Keep in mind that the tendon itself is not always where the primary defect is, and the mechanisms of the problem are not fully understood.

Treatment for Contracted Tendons

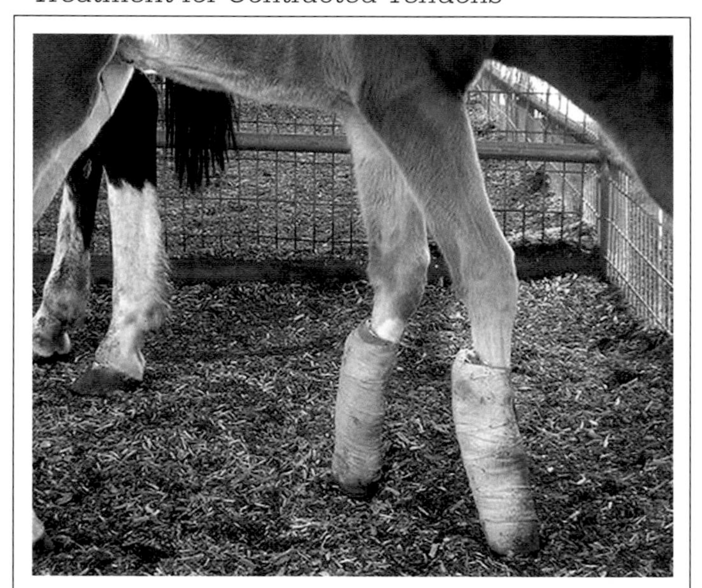

- Splinting is often used to treat foals with congenital flexural deformities.
- The splints help the flexor units to relax, but it's imperative that the splints be properly applied and monitored, so always work closely with your vet.
- For foals with acquired flexural deformities, changes in diet, exercise, and trimming and shoeing should be addressed to correct the problem.
- Surgical intervention is usually only considered after more conservative treatment methods fail.

can result in joint swelling and cysts or lead to osteochondritis dissecans (OCD), where small pieces of cartilage or bone are loose in a joint.

DODs can be caused by diet, rapid growth, trauma/injury, or possibly genetics. Diet is considered to be a major issue. The main culprits are likely excess grain as well as mineral imbalances. Calcium and phosphorus must be fed in the proper ratio. In addition, growing horses shouldn't receive an excess or deficiency of other minerals, including zinc, copper, selenium, iodine, fluoride, or heavy metals. An excess of zinc and deficiency in copper may be one aspect contributing to DODs. The only way to know for sure how much of these minerals a horse is getting is to have a complete analysis performed on all hay, pastures, and water, and look at the amounts contained in concentrates.

Owners should work with their vet to determine what to feed growing horses and the right amount for each stage of development. If a DOD is suspected in a foal, prompt veterinary intervention is needed. Treatment depends on the condition, and some cases require surgery.

Angular Deformity

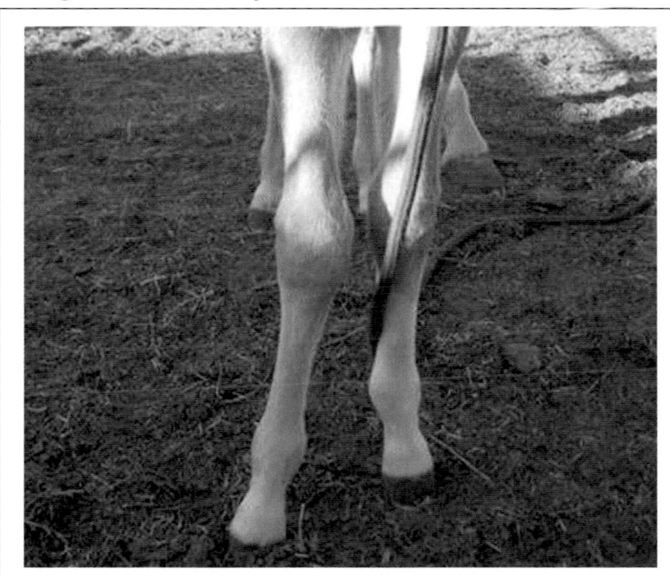

- Although there are other causes, angular limb deformities often occur when one part of the developing horse's bone grows faster than the other.
- If you suspect a deformity, don't wait to see if the foal grows out of it. Because the growth plates close up, prompt treatment is necessary.
- During treatment, the vet will likely recommend keeping the foal and dam confined to a small pen or stall, since foals are notoriously active and can easily cause further damage or injury.

Epiphysitis/Physitis

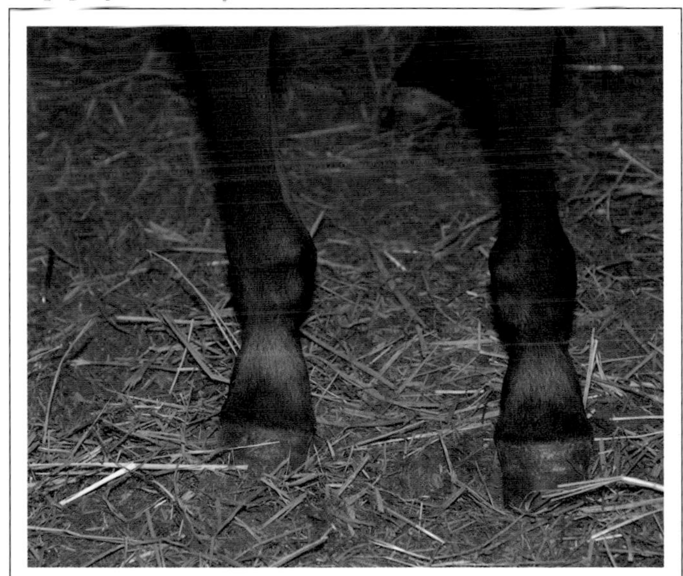

- Epiphysitis is considered a DOD and is also known as physitis or physeal dysplasia.
- Epiphysitis can occur in foals three to six months old at the ends of the cannon bones, creating pain and swelling around the ankle (shown in the photograph above).
- The problem can also occur in yearlings at the ends of the tibia or radius.
- Prompt treatment—mainly diet modifications—usually produces a positive outcome and can prevent permanent damage.

STANDING WRAPS

Properly applied standing wraps serve a variety of purposes

Standing wraps are fairly standard in the horse world. They are used for nonriding protection, support (as when a leg is "puffy" for various reasons), as bandage wraps over wounds, or over liniments, poultices, or sweats. As the basis for all these important functions, it's a good idea to know how to properly apply a standing wrap.

You don't want the wrap to create any pressure points. Applying wraps too tight, too loose, with wrinkles, with insufficient padding underneath, or leaving them on too long can all damage a horse's legs. For these reasons practice makes perfect. If it's your first time applying a wrap, it's a good idea to have an experienced horseperson check your work.

First, be sure your horse's legs are clean and dry (not counting liniments, poultices, or sweats). Then you'll need the

Standing Wrap 1

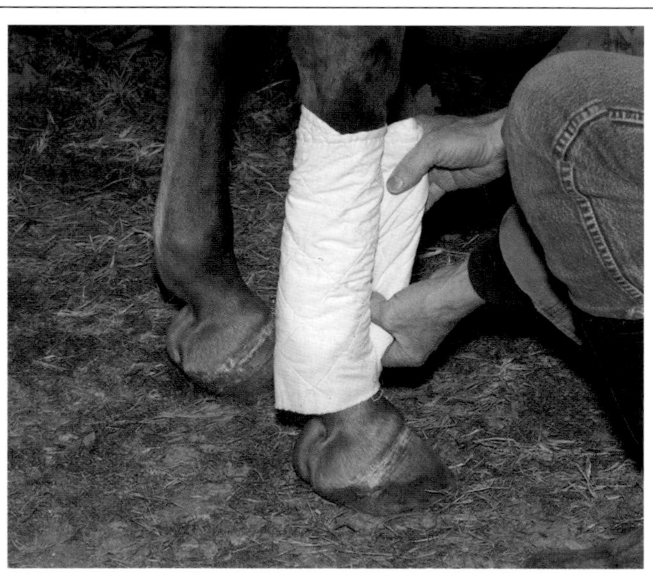

- Place the padding below the knee or hock, where it won't interfere with movement.
- Start the padding on the side of the horse's cannon bone so that it isn't placed over a tendon.
- Hold the padding in place with one hand until the first pass around the leg is complete.
- See page 60 for an explanation of how to apply polo wraps, which is similar to applying the stable bandages that go over the padding.

Standing Wrap 2

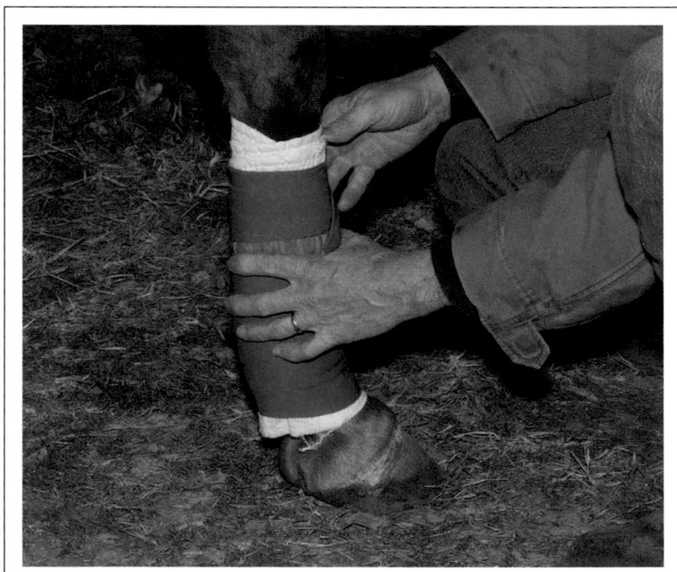

- Start your stable bandage below the upper edge of the padding or in the middle, depending on the horse's leg size and the length of the wrap.
- Wrap down around the fetlock area before working back up, overlapping each layer by 50 percent.
- Avoid any wrinkles and apply even tension. You should be able to insert a finger under the finished wrap.
- A 1/2-inch or so of padding should be visible above and below the finished standing wrap.

proper equipment. Since standing wraps are usually left on for several hours or overnight, you don't want too much pressure on the legs. That's where padding comes in. There are many types of padding now available. Quilts are usually made of cotton, polyester, or blends that contain some fill. Pillow wraps or "no bow" pads tend to be thicker and may not need to be wrapped around as many times as quilts. Whichever you choose, when the wrap is complete, there should be about a 1-inch thickness of padding. Quilts and pads come in different heights from 12 to 18 inches (and smaller pony sizes) and lengths (30 inches or more) depending on the size of the horse's leg.

The stable bandages that go over the padding can have some give but shouldn't be too elastic. They are generally 4 to 5 inches wide (slightly wider than exercise wraps) and come in different lengths.

Debris can get under the padding, and wraps may change position, creating pressure points, so remove or reapply standing wraps daily.

Bandage Wrap 1

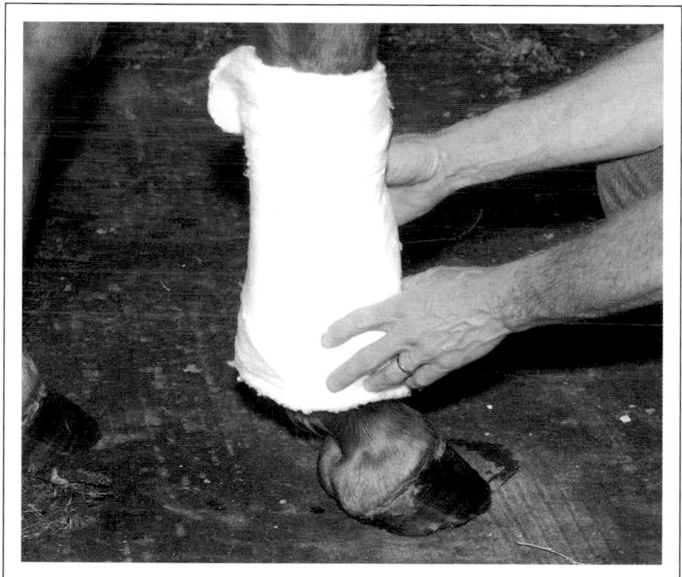

- For wounds, disposable cotton sheet padding (batting) is often used in place of reusable padding, and it's a good idea to keep a roll in your first-aid kit.
- Disposable cotton batting can be layered several times for adequate padding and folded to provide an ideal fit for trickier areas.
- It's important to avoid any wrinkles or clumps of material that will create a pressure point in the wrap.
- All wraps should be checked regularly and removed/reapplied daily.

Bandage Wrap 2

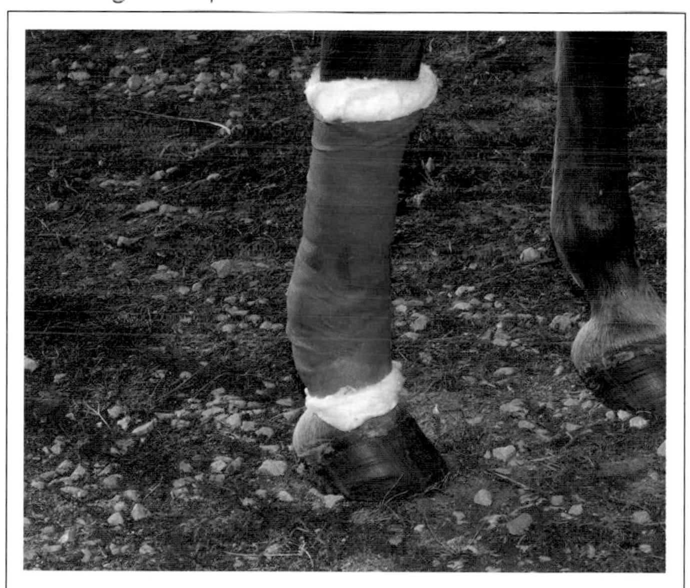

- A finished bandage wrap looks similar to a standing wrap, with padding visible above and below the wrap.
- Instead of using a stable bandage for the outer layer, disposable self-adhesive bandaging tape is often used for wound wraps.
- Keep a roll of self-adhesive bandaging tape like Vetrap in your first-aid kit.
- Appling wound bandages to other parts of the horse's leg can be tricky, but the same principles and materials usually apply.

HEAT AND COLD THERAPY

Learn how and when to apply heat and cold therapy

Just as with humans, heat and cold both have therapeutic effects in horses when used correctly and at the right time. Before using hot or cold therapy over an open wound, contact your vet. Cold or cryotherapy is best for acute injuries during the forty-eight hours following trauma. The inflammatory cycle can cause damage; cold therapy helps reduce blood flow to the injury and slow the migration of inflammatory cells, reducing inflammation and the swelling that accompanies it. Cold therapy also reduces pain. It should be applied for at least thirty minutes, and can be reapplied every four hours. For cold therapy to be most effective, skin temperatures must be lowered sufficiently, but if they're lowered too much, tissue damage can occur. The same is true of heat but in the opposite direction—too little heat will be ineffective, but applying something that's hotter than 130 degrees directly to the skin can cause tissue damage.

Boots

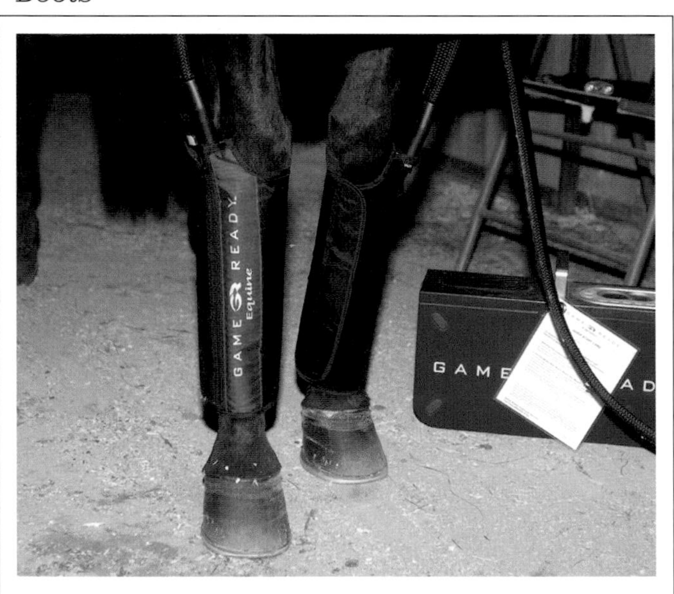

- There are a variety of boots and wraps available commercially for applying cold therapy to a horse's lower leg.
- These can be convenient if you are treating a chronic problem, such as an old tendon issue, or if you like to use ice after workouts as a preventative measure.
- Some specialty boots can also be used for heat therapy.
- Follow manufacturer instructions to ensure safe and effective application.

Bucket Application

- A bucket filled with ice and water can be an effective way to apply cold therapy to a horse's lower leg—use for thirty minutes at a time.
- A bucket can also be filled with hot water for heat therapy, but stick your hand in and be sure you can tolerate the temperature.
- Using a bucket takes careful monitoring and a patient horse.
- Use a textured rubber bucket (like the one shown above) or apply nonslip grips to the bottom of the bucket.

Heat therapy is used after the acute stage when swelling has subsided or stabilized—generally two or three days after the trauma. Heat will increase blood flow, which you don't want during the acute phase. Using heat later in the injury's healing or for chronic problems will improve circulation to bring nutrients to the area, aid in tissue oxygenation, encourage reabsorption of excess fluids, and help remove waste products. And unlike cold, which can make tissues less elastic, heat helps relax muscles and soft tissues. Heat also decreases pain, sometimes with lasting benefits.

Reusable commercial heat and cold packs intended for humans can be used on horses, but choose packs that can conform to the horse's leg. Place a piece of cloth between the horse and the pack. You can apply the pack by hand, or depending on the site, use a stable wrap to gently hold it in place.

Hosing

- Hot water hosing is another way to deliver heat therapy.
- With warm water, be sure it's not too hot by running it over your hand.
- Cold water often isn't cold enough to lower the skin's temperature sufficiently for cold therapy.
- Towels can also be wet with hot water and applied to the horse or used under a heat pack—just be sure they're not too hot.

Magnet Therapy

- There are magnetic boots for virtually every part of a horse's leg.
- Supporters of magnet therapy claim that magnets have therapeutic effects in horses, humans, and other animals.
- Claims that magnets have therapeutic benefits have not been backed up by scientific evidence.
- While magnet therapy likely won't do any harm, it should not take the place of proven methods, such as hot and cold therapy.

POULTICES

Poultices are sometimes used to reduce inflammation or provide cooling effects

The word *poultice* comes from the Latin word for porridge, and indeed poultices are about the same consistency as porridge. Poultices are used for different reasons based on their ingredients. Ingredients can include natural clays, Epsom salt, cooling peppermint or menthol, or medications to help prevent infection. Common uses for poultices include providing cooling benefits and drawing out swelling in the lower leg, or drawing out abscesses in the hoof.

Many top riders and trainers poultice their horses as a preventative measure after workouts. If you plan to apply a poultice as a preventative measure after a workout or competition per your trainer's advice, it's generally safe to do so. Just

Poultice Application 1

- Poultices are often applied to the lower leg between the bottom of the knee and the fetlock.
- They can be applied by hand, but you may want to use a rubber glove.
- An alternative is to apply the poultice material directly to the paper covering (see step 2).
- Read the section of this chapter on applying standing wraps (page 188) before attempting to apply a poultice wrap.

Poultice Application 2

- Many poultices are designed to have a cooling effect, so paper wetted with cool water is applied over the poultice material and will easily stick to it.
- The wetted paper helps keep the leg cool and works as a barrier between the poultice and the padding.
- You can buy poultice paper online or at a tack store, or use butcher paper or a paper lunch sack cut open.

be sure you follow manufacturer instructions on commercial poultices, and test the product on a small area of skin first to be sure it doesn't cause irritation.

Do not apply poultices over open wounds. Consult your vet before using a poultice for an injury or undiagnosed condition causing swelling. After the acute phase, the vet may recommend a poultice to help reduce swelling. For example, a poultice with Epsom salts and menthol may be beneficial, as the Epsom salts can help draw out fluid, while the menthol increases circulation.

The farrier or vet may also recommend a poultice to treat a hoof abscess. If the affected hoof has a shoe, the shoe will need to be removed first. There are commercial medicated poultices for hooves, or there are homemade ones often containing Epsom salts and bran. The farrier or vet can advise what type of poultice to use. The poultice is held in place over the hoof using a wrap and hoof boot, or just a wrap. For instructions on applying a basic hoof wrap, see page 218.

Poultice Application 3

- As with any wrap that will be left on for several hours or overnight, it's important to apply adequate padding.
- Disposable cotton batting can be used for padding as it would with a bandage wrap, or you can use washable padding suited for standing wraps.
- Never apply poultices or other wraps to a dirty leg.
- Be sure you avoid bunching or wrinkles.

Poultice Application 4

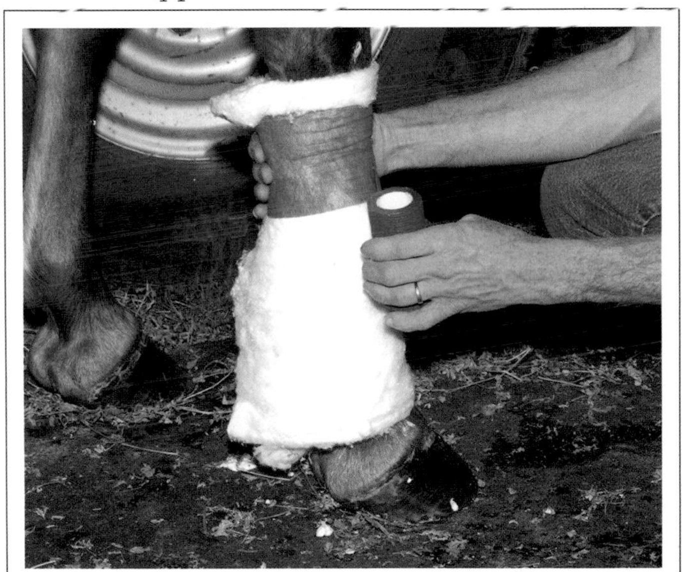

- It's important to use breathable materials when applying a poultice wrap to a horse's leg.
- The outer layer can be self-adhesive disposable vet tape or a washable stable bandage.
- As with other wraps, keep horses in a clean stall or small pen, not turned out, while they're wearing poultice wraps.
- Poultice wraps should be removed after twelve hours (or reapplied if recommended by the vet).

LINIMENTS

Liniments are often applied as analgesics to ease minor aches and pains

Like poultices, liniments vary based on their ingredients, with some designed to provide cooling effects, others heating effects. Certain liniments also claim to reduce swelling or improve blood flow. Most liniments are created as analgesics similar to topical human products, which are rubbed on to help relieve minor aches, pains, and muscle soreness.

Many liniments contain herbs, botanicals, and oils, such as menthol, eucalyptus, capsaicin, camphor, arnica, and mint oils. Analgesics should not be used over open wounds. Some liniments contain antiseptics and are labeled as safe for minor wounds. As with poultices, don't use liniments to treat an injury or swelling without first consulting your veterinarian.

Liniments

- Liniments are a broad category of products, which can contain a variety of ingredients.
- Liniments commonly come in gels, liquids, and washes.
- Read the ingredients, uses, and directions before deciding on a liniment.
- If you want to try a liniment but still aren't sure which one to use on your horse, ask a trusted horse trainer or veterinarian for a recommendation.

Liniment Washes

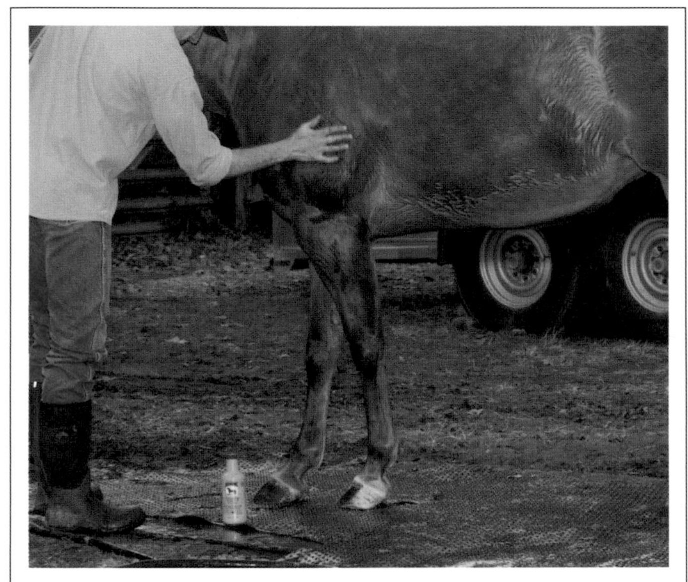

- Some liniments are designed as washes to be used after a workout.
- Liniment washes may contain a bit of soap as well as traditional cooling, bracing, or analgesic ingredients.
- Certain washes should be diluted, so read the instructions carefully before application.
- If the liniment you use appears to irritate your horse's skin, discontinue application of the product and thoroughly wash away any residue.

Liniments can be applied in a variety of ways depending on the horse owner's preference and the type of liniment. Some liniments are used before work to help warm up the horse's soft tissues. However, these should never take the place of a warm-up under saddle. Skipping a proper warm-up can lead to injury.

Other liniments are designed to be rubbed on after a workout. Some are massaged on and left, others are applied under standing wraps overnight, and others are diluted or used full strength as rinses or washes. Be sure to follow manufacturer instructions. It's also a good idea to test the liniment on a small area of your horse's skin and leave it there for the specified amount of time to ensure it doesn't cause irritation before using it on all four legs or over the horse's entire body.

If you compete and use liniments regularly or before competition, be sure they don't contain ingredients that are against the rules or will make your horse test positive for banned substances. Remember that products applied topically can enter the horse's system.

Application

- Most liniments are designed to be rubbed on by hand.
- The massaging involved in applying the liniments may also have a beneficial effect on muscles.
- If you use liniments to help warm your horse up before riding, it's usually a good idea to rinse them off when you're done, depending on manufacturer instructions.
- Liniments can be applied using rubber gloves if you don't want to feel the effects on your hands.

Setting Up

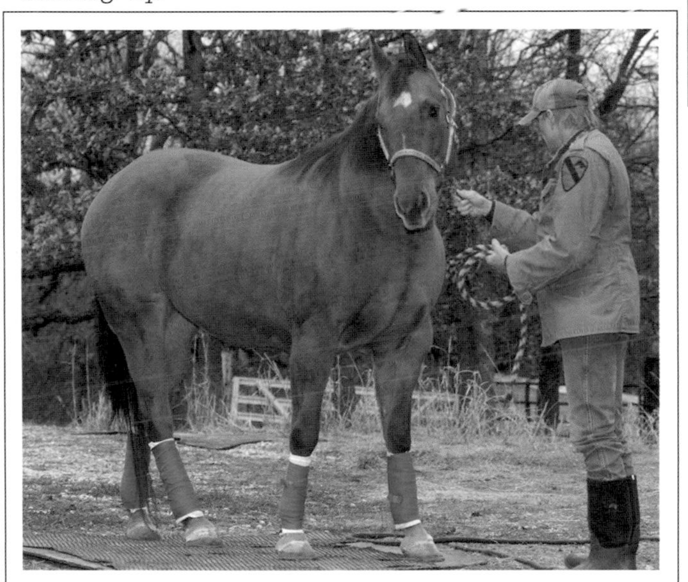

- Applying a liniment under standing wraps is called setting up, and the wraps are usually left on overnight. Check your liniment's packaging to be sure it's labeled for use under wraps.
- See page 188 on how to properly apply standing wraps, as wrapping incorrectly can cause injury.
- Remember that liniments shouldn't be used in place of proper veterinary care.
- Always check with your vet before applying liniments to a swollen leg, and never use liniments on a fresh trauma.

SWEATS AND DMSO

Sweat wraps and DMSO may be used alone or in combination to treat inflammation

Sweat wraps are designed to reduce fluid build up in the lower leg by "sweating it out." They should not be used over open wounds or on recent injuries, and owners should check with the vet before using a sweat wrap.

Sweat wraps involve slathering the lower leg from below the knee to the fetlock with a product such as nitrofurazone ointment, mineral oil, glycerin, petroleum jelly, or DMSO. The veterinarian can recommend which product is best for the particular situation. The product of choice is then covered with a layer of lightweight plastic wrap, which can be purchased at any grocery store. Then the leg is wrapped in a standing wrap.

Products

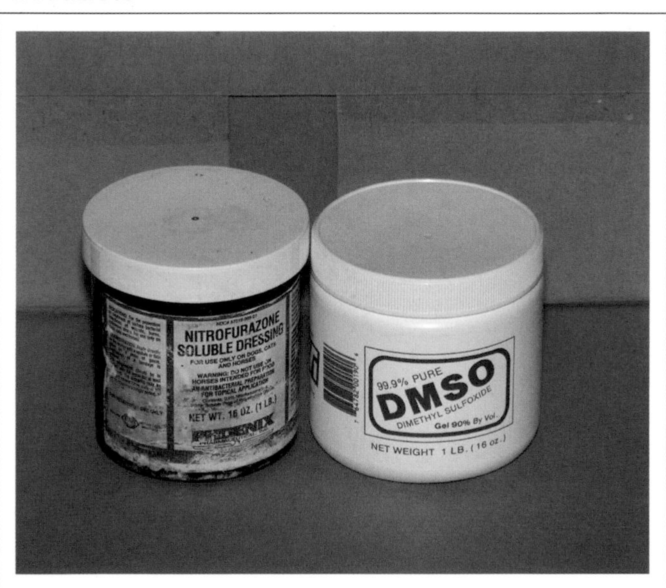

- DMSO may be used as a topical anti-inflammatory under sweat wraps or standing wraps.
- Nitrofurazone dressing is another commonly used product for sweat wraps.
- Nitrofurazone is a topical antibacterial also sold under the brand names Fura-Zone and Furacin.
- Both DMSO and nitrofurazone can be kept on hand in your first-aid kit if your veterinarian recommends.

DMSO Application

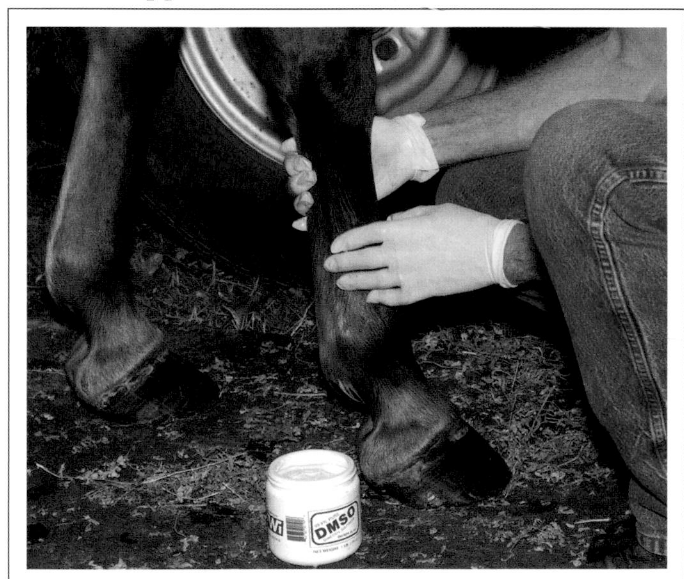

- DMSO is available in liquid or gel forms and is known for its foul smell.
- DMSO has a long-standing history of use in horses and has been approved by the FDA since the 1970s.
- It's extremely important to wear rubber gloves when handling or applying DMSO, as it will be absorbed by your skin.
- Be sure your horse's leg is dry and clean of all previously used substances, including fly spray or shampoo residue before you apply DMSO.

DMSO (dimethylsulfoxide) is a solvent that originates as a byproduct from the wood-processing industry and has been in popular use for several decades. DMSO is commonly used as a topical anti-inflammatory and is sometimes used under sweat wraps or standing wraps.

By itself DMSO is nontoxic, but is readily absorbed and will take along with it other substances. This can be good for delivery of certain substances, such as other anti-inflammatories or antibiotics per veterinary advice, but care must be taken not to accidentally mix DMSO with other substances. For this reason, it's important to only apply DMSO on clean skin that's free of all other substances (except those recommended by the vet).

DMSO can have a heating action, so it is usually not recommended for injuries during the acute stage. It generally should also not be used over open wounds. DMSO is often used as part of the treatment plan for tendon and ligament injuries, and other traumas that involve swelling; however, only use DMSO per veterinary advice.

Sweat Wrap Application 1

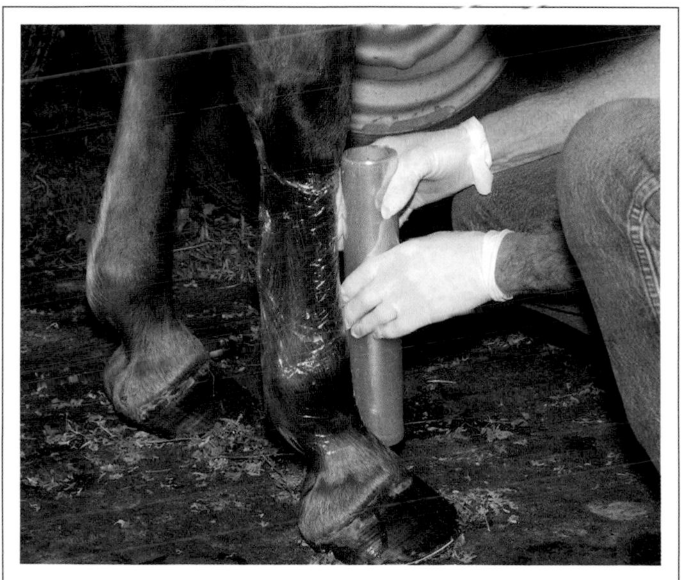

- To apply a sweat wrap, the topical product of choice should be liberally slathered onto the leg where the wrap will go (if using DMSO follow product directions or veterinary instructions for application).
- The plastic wrap should be placed around the leg over the same area as the product, with care taken to keep it smooth.
- After the plastic wrap is in place, application is the same as for a standing wrap, described on page 188.

Sweat Wrap Application 2

- Disposable cotton batting or washable standing wrap quilts can be used for padding over the plastic wrap.
- As with a standing wrap, use at least a 1-inch thickness of padding.
- The padding can then be wrapped over with self-adhesive disposable vet tape or a stable bandage.
- Sweat wraps should not be left on longer than twelve hours, and notify your veterinarian if the leg looks irritated or the condition worsens.

PAIN MEDICATIONS

Nonsteroidal anti-inflammatory drugs are the front line in pain relief

When we have aches or pains, we may take aspirin, naproxen, or ibuprofen. These fall into the drug category called nonsteroidal anti-inflammatory drugs, abbreviated as NSAIDs, which are mostly perscription drugs. For horses, NSAIDs include phenylbutazone or "bute," flunixin meglumine (Banamine), ketoprofen (Ketofen), meclofenamic acid (Arquel), aspirin, and naproxen.

NSAIDs can mask pain and fever, so it's important not to administer an NSAID for a purpose other than that for which it was prescribed without consulting the vet. NSAIDs inhibit the body's production of prostaglandins and other inflammatory mediators, thus reducing swelling, pain, and heat. However, they can also inhibit enzymes that protect the gastrointestinal tract.

Common NSAID side effects include gastrointestinal ulcers, colitis (colon problems), and kidney damage. Therefore, they should be used at the lowest dose needed and for the short-

Bute Tablets

- Bute (phenylbutazone) is available in many forms, but one-gram tablets are the most common.
- Tablets can be crushed or dissolved in water and added to sweet feed.
- Dosage will be determined by the veterinarian, but two grams administered twice a day for five days is usually the maximum for a full-size horse, or two grams a day for long-term use.
- The higher the dosage and the longer bute is used, the more risk there is for developing side effects.

Bute Powder

- Bute is currently the most common NSAID for long-term use.
- Bute is often prescribed for injuries, laminitis, and chronic pain from joint disease or osteoarthritis.
- Previously, owners have tried many things to get their horses to eat bute, including adding Kool Aid powder to the bute and sweet feed.
- Nowadays, there are flavored, powdered versions of bute available that are more palatable (shown above).

est amount of time. The side effects are the same whether the NSAID is administered orally or via injection.

When your horse is on NSAIDs, monitor him for signs of gastrointestinal problems, which can include colic, diarrhea, lack of appetite, depression, fever, grinding teeth, or a decrease in manure production. Signs of kidney problems include excess urine or dark colored urine. If any of the above symptoms are present, contact the veterinarian.

Proper hydration can help reduce the risks, so be sure your horse has constant, easy access to clean water. Frequent feedings, plenty of forage, and linoleic acid (for example, adding safflower or corn oil to the horse's feed) may also reduce a horse's risks for gastrointestinal problems, so consult your vet regarding diet modifications.

If NSAIDs like bute must be given long term to manage chronic pain, talk to your vet about the possible benefits of giving the horse occasional "vacations" from bute use. Another option is the new COX-2 inhibitor firocoxib (EQUIOXX), which has fewer gastrointestinal side effects than other NSAIDs (other NSAIDs inhibit COX-1 and COX-2 enzymes).

Bute Paste

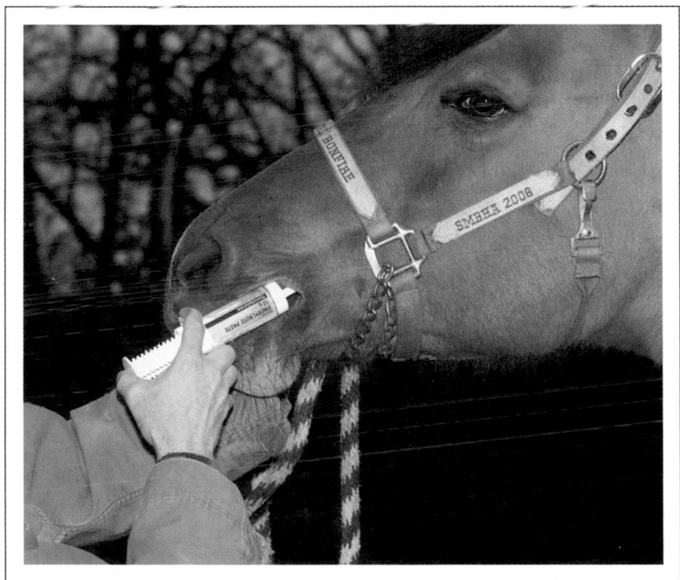

- In addition to oral tablets and powders, bute is also available in a paste syringe.
- However, horses quickly learn that the taste is not pleasant and will resist the syringe, so it's best to only use syringes for emergency or short-term dosages.
- Only give bute at the recommended dosage for the prescribed time.
- Do not give bute to a horse that it wasn't prescribed for without first consulting your veterinarian.

Injection

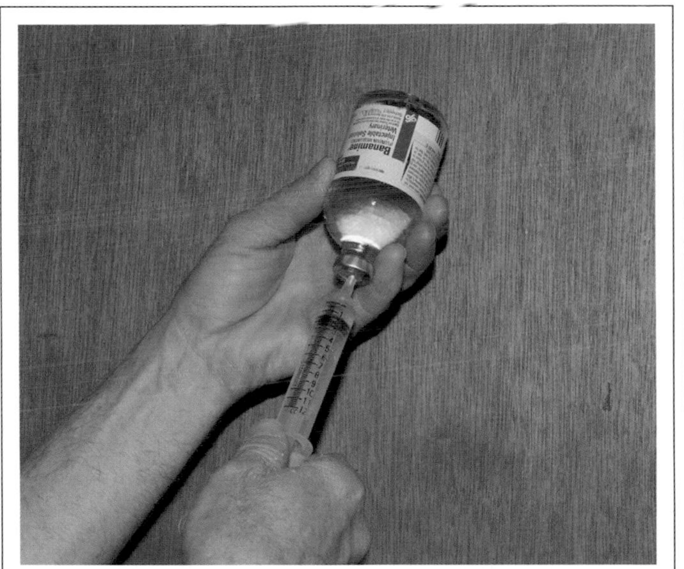

- Most NSAIDs are available via injection as well as orally. The vet may inject a dosage for rapid relief and then prescribe additional oral doses.
- Flunixin meglumine (Banamine) can be given intravenously, intramuscularly, or in a paste or granule form.
- In addition to all the oral forms of bute, it can also be given intravenously short term.
- Ketoprofen can also be given intravenously or orally.

ARTHROSCOPIC SURGERY

Arthroscopy and arthroscopic surgery are valuable tools for treating and visualizing what's going on inside a joint

Arthroscopy means to look inside a joint, and many people are familiar with arthroscopic surgery, which is commonly performed on the knees of human athletes. Arthroscopy and arthroscopic surgery have similar benefits for horses and are used to diagnose joint issues and to remove bone chips and other irritants within the joint.

Arthroscopic surgery involves inserting a pencil-sized camera into the joint via a small incision and making one or two additional incisions for any additional surgical instruments needed. The arthroscope transmits images to a television screen to guide the surgeon.

Because arthroscopic surgery is minimally invasive, it carries

Diagnostic Arthroscopy

- The arthroscope used during arthroscopy and arthroscopic surgery allows veterinarians to clearly visualize what's going on within the joint.
- An arthroscope is only 4 millimeters in diameter, so it can be inserted through only a small stab incision.
- Arthroscopy can be used prior to other types of surgeries for a complete and accurate diagnosis.
- Although x-ray is the frontline diagnostic, arthroscopy can be used on joints like the stifle's femorotibial joint, where x-ray and ultrasound fail. .

Arthroscopic Surgery Common Uses

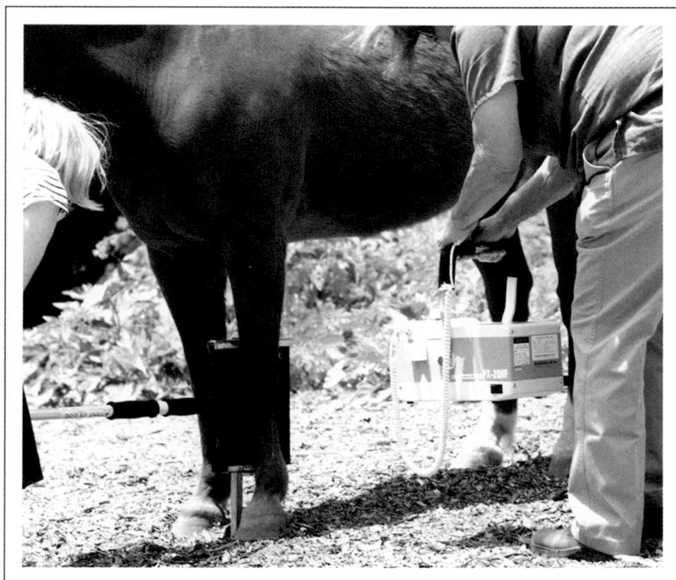

- While x-rays can show bone damage, they usually do not provide enough information regarding damage to articular cartilage, which is where procedures like arthroscopy and arthroscopic surgery can help.
- Intra-articular chip fractures are one of the most common uses for arthroscopic surgery in mature horses.
- Intra-articular fractures cross into the joint surface and cause cartilage damage.
- In younger animals, arthroscopic surgery is often used to treat osteochondritis dissecans.

less risk than most surgeries. The horse has to undergo general anesthesia, and all surgery carries some risk for infection, but these risks are both minimal with arthroscopy.

During arthroscopic surgery, the joint is inspected, and loose cartilage or bone chips can be removed, the edges of lesions cleaned, and the joint flushed. Removing irritants can be very beneficial to a horse; however, arthroscopic surgery is not a cure for degenerative joint disease (DJD) and can't rejuvenate thinning articular cartilage. If nothing surgical can be done within the joint, arthroscopy at least gives the vet an idea of how to create the best treatment or maintenance plan.

After the procedure, the surgical incisions usually can be closed with just one suture each and are unlikely to leave obvious scars. Most horses are quite unbothered by the surgery and are able to return home the next day. Stall rest will likely be recommended for two weeks following the procedure, and additional aftercare may be needed depending on the horse, the affected joint, and what took place during the surgery. However, recovery time is generally much shorter than that expected after more invasive surgeries.

Benefits and Drawbacks

- Used to diagnose and treat joint issues, arthroscopy and arthroscopic surgery are breakthrough diagnostic and surgical techniques that crossed over from human medicine to begin helping horses in the 1970s.
- Like human athletes, horses can often compete just months after surgery.
- While arthroscopy and arthroscopic surgery are invaluable, the main drawback is the cost.
- As always, discuss total cost estimates and possible outcomes with your vet before scheduling any procedure.

Stall Rest

- Stall rest will likely be recommended for two weeks following arthroscopic surgery.
- Other than stall rest, aftercare will vary based on the horse, which joint was affected, and what exactly took place during the surgery.
- However, because arthroscopic surgery is minimally invasive, recovery time after surgery is generally much shorter than that expected after more invasive surgeries.

JOINT INJECTIONS

Intra-articular, intramuscular, and intravenous injections can help prevent and treat joint disease

Degenerative joint disease (DJD), which may lead to the bone changes associated with osteoarthritis, affects almost every horse eventually and is likely the most common career-ending condition (see pages 80–92). Today's arsenal of joint injections can help keep joints healthy and treat the destructive cycle of inflammation associated with joint disease.

When treating a joint problem, first get an accurate diagnosis. Then work with your vet to determine if injections are an option and if so which to use and how often.

Hyaluronan, sodium hyaluronate, or hyaluronic acid are all commonly abbreviated as HA. HA is naturally produced by the synovial membrane that helps lubricate the cartilage and

Infection

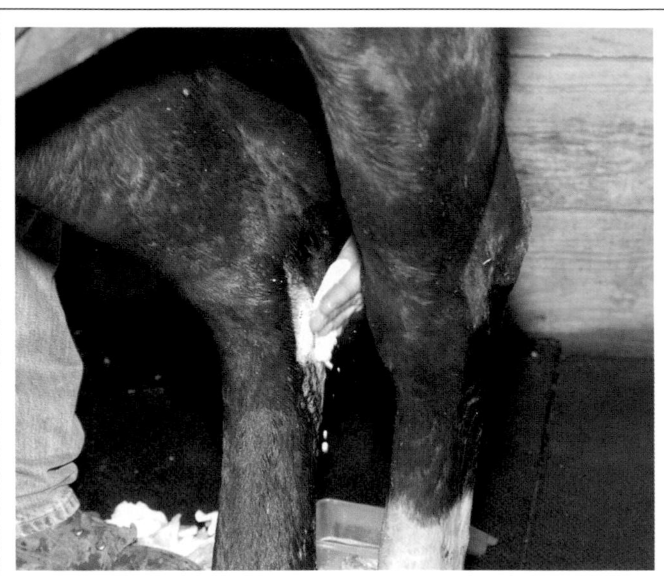

- When the vet uses sterile techniques and proper preparation (shown in the photograph above), there is minimal risk for infection following a joint injection.
- If you notice any signs of infection after an injection—heat, tenderness, or swelling around the joint—call the vet immediately.
- In addition to infection, swelling or sensitivity around the injection site may also indicate an adverse reaction to the substance, and these symptoms require a vet consult.

Intra-articular Injections 1

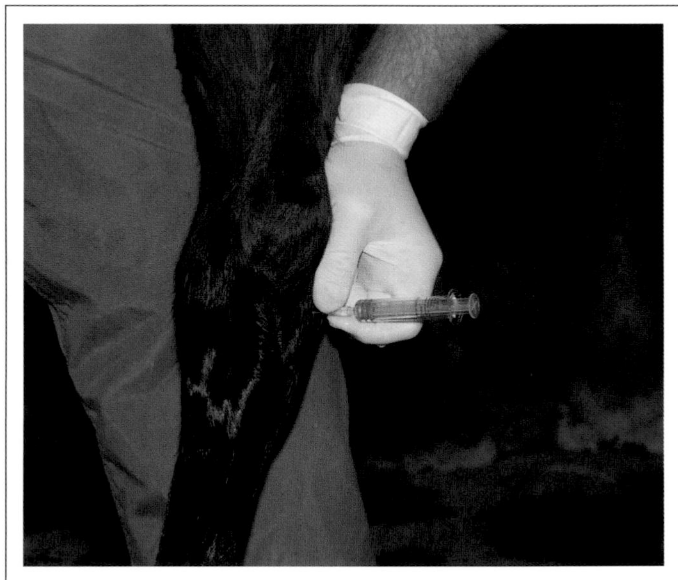

- The most commonly injected joints are the fetlock, coffin, pastern, and stifle joints, along with particular knee and hock joints.
- The vet may recommend more than one injection or product.
- For example, HA may be given as an intra-articular injection at the same time that PSGAG is given as an intramuscular injection.
- HA is often combined with a corticosteroid and given as an intra-articular injection, which is called "white acid" for its color.

inhibit certain inflammatory mediators that can damage the cartilage. When a horse has early signs of joint trouble, such as synovitis and capsulitis (see page 96), an injection of HA can help stimulate natural HA production and improve the quality of joint fluid/lubrication. Legend is the most common brand name of HA and is available for intra-articular (within the joint) or intravenous injection.

Polysulfated glycosaminoglycan (PSGAG) is often used to help prevent cartilage degeneration due to its apparent ability to inhibit destructive enzymes. Adequan is the common brand name associated with PSGAG injections and is available in two forms: Adequan I.M. is given intramuscularly to help prevent or reduce the cycle of degeneration within joints, and Adequan I.A. is for intra-articular use.

When used correctly, corticosteroids are often a beneficial intra-articular anti-inflammatory. The inflammatory cycle is what causes continued degeneration within the joint, and corticosteroids help inhibit inflammatory enzymes. By reducing inflammation, corticosteroids can also reduce the associated heat, pain, and lameness.

Intra-articular Injections 2

- After a joint injection, a day of stall rest followed by two days of turnout or hand walking may precede light work under saddle.
- If joint disease or a joint injury is present, it's important to address the cause and not just treat the symptoms with injections. A change in career, workload, or footing may be needed.
- Injections like corticosteroids can make the horse feel better, so care must be taken to avoid further joint damage, injury, or re-injury.

Intravenous and Intramuscular Injections

- Legend (HA) is available as an intravenous injection, which may be used in combination with an intra-articular injection.
- Legend I.V. appears quite effective and has the benefit of reaching multiple joints (not just one, as with an intra-articular injection).
- Adequan I.M. is likely the most common PSGAG treatment and is given intramuscularly.
- Injections often must be repeated at regular intervals for maximum benefit, and your vet can help you determine an effective schedule.

ESWT AND IRAP

One shows promise for a number of musculoskeletal issues, while the other can help horses with joint disease

Just as new advances in human medicine are always on the horizon, vet schools and researchers are constantly pioneering new treatments for horses. While these cutting edge treatments may not be available to the average horse owner at first, they generally become more affordable and accessible after a period of proven success, and they can change the way certain injuries or illnesses are treated.

Extracorporeal shock wave therapy (ESWT) is where vets use a shock wave machine to aim high-pressure sound waves at an injured site within the body. ESWT is a relatively new therapy, and its uses, benefits, and guidelines are still being researched. ESWT appears to aid in the healing of a num-

ESWT

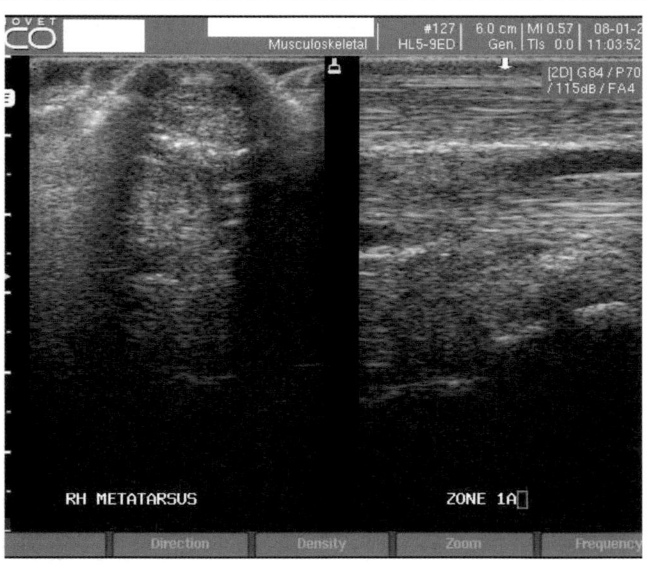

- This ultrasound shows a suspensory ligament avulsion lesion, which occurred as an acute injury during jumping.
- The twelve-year-old mare was treated with stall rest, NSAIDs, bandaging, and ESWT.
- The mare made a full recovery and is now jumping at her previous level.
- In addition to ESWT, there is radial pressure wave therapy (RPWT), which is sometimes used as an alternative to ESWT.

ESWT 2

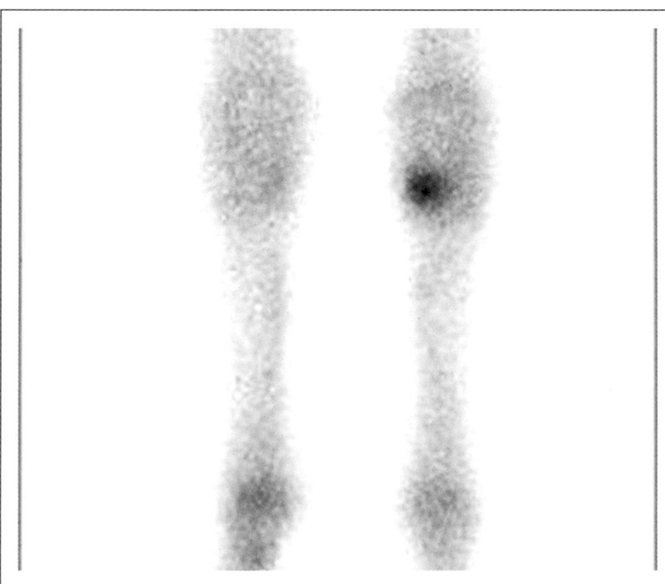

- This nuclear scintigraphy image shows where a carpal fracture occurred. The mare was treated with surgical debridement, stall rest, NSAIDs, and ESWT.
- Most conditions require multiple ESWT treatments over several weeks.
- The fracture healed as a non-union fracture of the third carpal bone, and the horse is pasture sound as a broodmare.
- Because ESWT can reduce pain for a few days after treatment, it's important not to let horses overdo it after the session.

ber of musculoskeletal conditions. Exactly how it aids healing isn't completely understood but seems to be due in part to ESWT's ability to improve blood flow and stimulate cells. ESWT can also stimulate bone formation and remodeling in certain situations. Studies show that ESWT speeds healing in many cases of tendinitis (tendon injury) and desmitis (ligament injury), especially injuries near the bone attachment sites. It's also demonstrating promise for treatment of stress fractures, navicular area pain, bone spavin, and other types of osteoarthritis and musculoskeletal conditions.

Interleukin-1 Receptor Antagonist Protein (IRAP) is an exciting new gene therapy for joint problems and degenerative joint disease (DJD) (see page 80 for information on DJD). Stopping the inflammatory cycle can halt the degeneration in a joint. The protein known as Interleukin-1 is a key player in the inflammatory and degenerative cycle. (See photos below for more information.)

IRAP 1

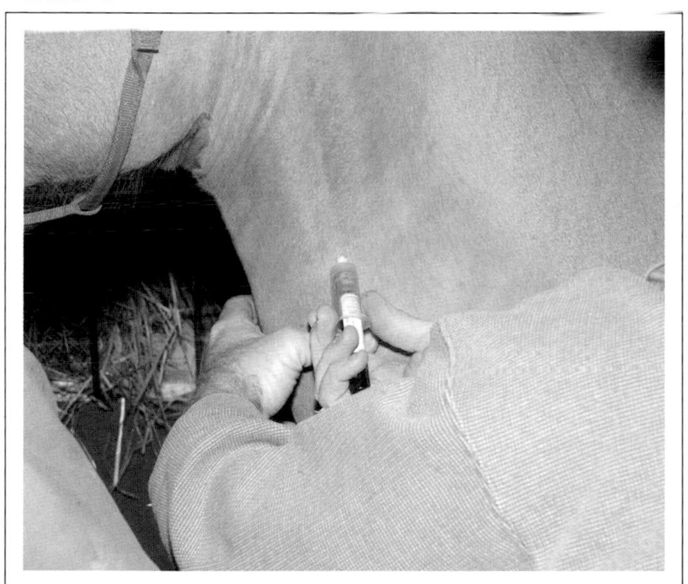

- IRAP therapy starts by taking a syringe of blood from the affected horse.
- The blood is then specially processed and incubated, and the beneficial fluids created are injected into the affected joint(s).
- A series of three to five injections are usually recommended.
- IRAP may produce long-term benefits for horses with DJD because it has the potential to halt or slow the damaging cycle within the joint.

IRAP 2

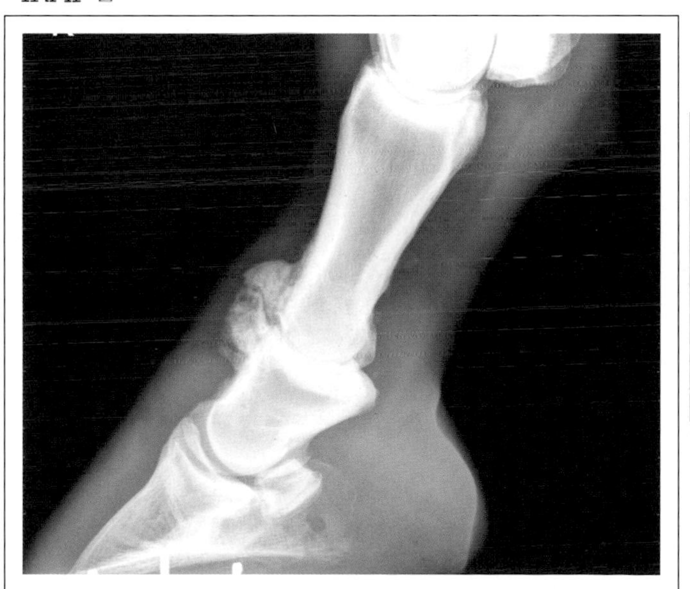

- This x-ray is of an eighteen-year-old horse with high ringbone, who was suffering from chronic lameness and was treated with NSAIDs and IRAP injections. The horse is now sound and comfortable for trail use.
- For IRAP therapy, a syringe of blood is taken from the horse and incubated with beads covered in substances that promote healing and encourage production of the antagonist protein.
- The beneficial fluids are then separated from the blood and injected into the affected joint to inhibit Interleukin-1.

STEM CELL AND ACELL

These therapies involve injection at the site of tendon and ligament injuries for better, faster healing

Stem cell and ACell, as well as platelet rich plasma therapy are among the cutting edge treatments focusing on tendon and ligament injuries, which commonly affect all types of horses. As use and research continues, expect these therapies to become more common for tendon and ligament issues as well as other musculoskeletal conditions.

Horse stem cell therapy doesn't carry the controversy that human stem cell therapy does, as equine stem cell therapy uses stem cells harvested from the injured horse's bone marrow or fat. A stem cell is a certain type of cell that can transform into a cell for a specific body part. For example, a sample of fat can be taken from a horse's hip area and sent to

Stem Cell Therapy

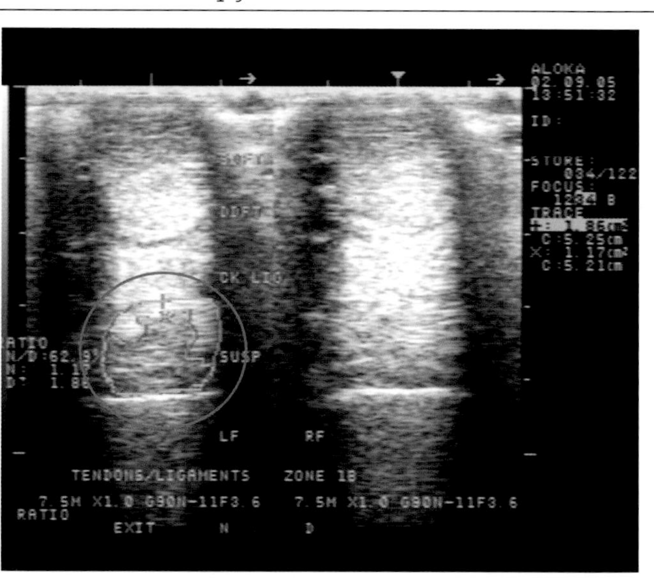

- This ultrasound shows a suspensory injury that occurred while the twelve-year-old horse was jumping.
- The gelding was treated with stall rest, NSAIDs, Vet-Stem stem cell injection, and extracorporeal shock-wave therapy. The horse made a full recovery.
- Stem cells harvested from fat don't have a long shelf live, and the lab usually has them ready for injection within forty-eight hours.
- In addition stem cells are attracted to cartilage damage and show promise for joints.

ACell Therapy Before

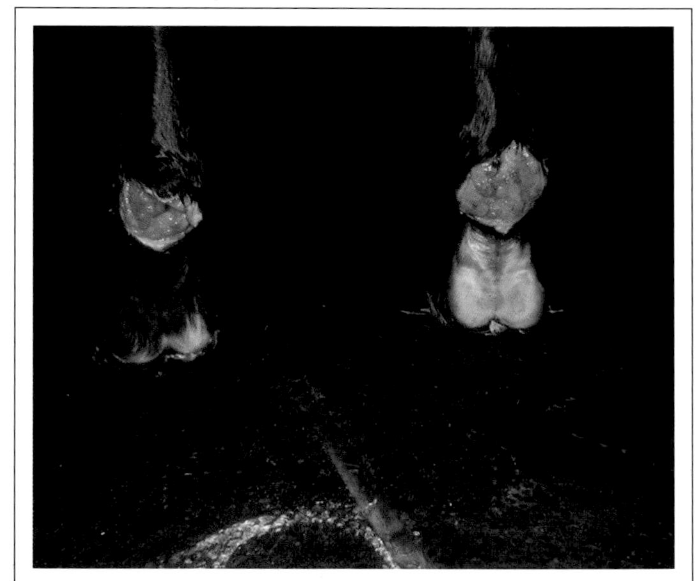

- This photo shows a four-year-old racehorse gelding with a history of "run-down" injuries. (Case history and photos provided from ACell, Inc.)
- A week before presentation this horse ran on a sloppy track. He presented with deep, full thickness subcutaneous tissue injuries on the back of both hind fetlocks (as seen in photo) and was treated with ACell Vet.
- The horse was discharged less than two weeks later, and the owner continued bandaging the legs at home.

a lab. The lab can recover the stem cells, which are in the fat sample along with other types of cells, and the stem cells can then be injected at the site of a tendon or ligament injury, where they can produce healthy new tissue rather than the weaker scar tissue the site may have produced during healing without stem cells.

In addition to tendon and ligament injuries, stem cell therapy also shows promise as a treatment for some bone and joint issues.

ACell therapy is tissue engineering using extracellular matrix material. The aims of treatment are similar to stem cell therapy, and like stem cell therapy, ACell is often used to treat tendon and ligament injuries. ACell therapy involves injecting an ACell mixture (derived from pig urinary bladder matrix) into the site of tendon or ligament injury to initiate healing and attract the body's own stem cells. ACell products are also promoted to help heal hoof wall injuries, wounds (including surgical sites), and burns.

ACell Therapy After

- The four-year-old racehorse shown in the before photo made a full recovery and was able to return to a successful racing career.
- Controlled exercise and a gradual return to work are needed as part of the recovery.
- ACell therapy aims to improve healing after an injury and reduce recovery time.
- ACell is used after initial swelling has gone down and location of the injury and a diagnosis have been determined.

Platelet Rich Plasma Therapy

- Platelet rich plasma therapy involves injecting a platelet-rich by-product of the patient's blood at the site of injury.
- The therapy can improve healing at the cellular level in tendon and ligament injuries.
- Because platelet rich plasma therapy stimulates inflammation, it may help in chronic tendonitis.
- If you're interested in a new treatment, talk to your vet. Some cutting edge treatments are only available through select veterinarians, universities, or equine hospitals.

COMPLEMENTARY THERAPIES

Complementary or alternative therapies can work with traditional treatments to reduce pain and aid healing

Complementary therapies include chiropractic, acupuncture, acupressure, homeopathy, and massage. These treatments may benefit your horse when used in conjunction with veterinary care, but should not be a replacement for traditional veterinary care.

If your horse is dealing with pain or lameness, and you're interested in combining complementary therapy with the treatment your horse is already getting, ask your regular veterinarian if this is advisable given the specific condition. Complementary therapies can also be used as a preventative measure. Either way, work with your trusted veterinarian to decide what's best for your particular horse.

Chiropractic

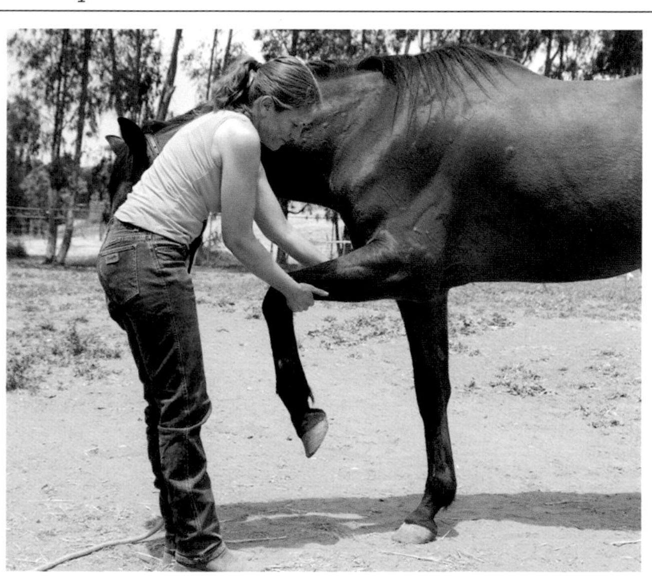

- Chiropractic adjustments are usually made using applied force via the hands and arms.
- Many performance horses are now treated with regular chiropractic adjustments to augment traditional veterinary care.
- One of the main theories behind chiropractic care is that proper structure and function of the musculoskeletal system leads to improved health.
- Chiropractic care often focuses on the spine and aims to correct any mechanical and musculoskeletal issues.

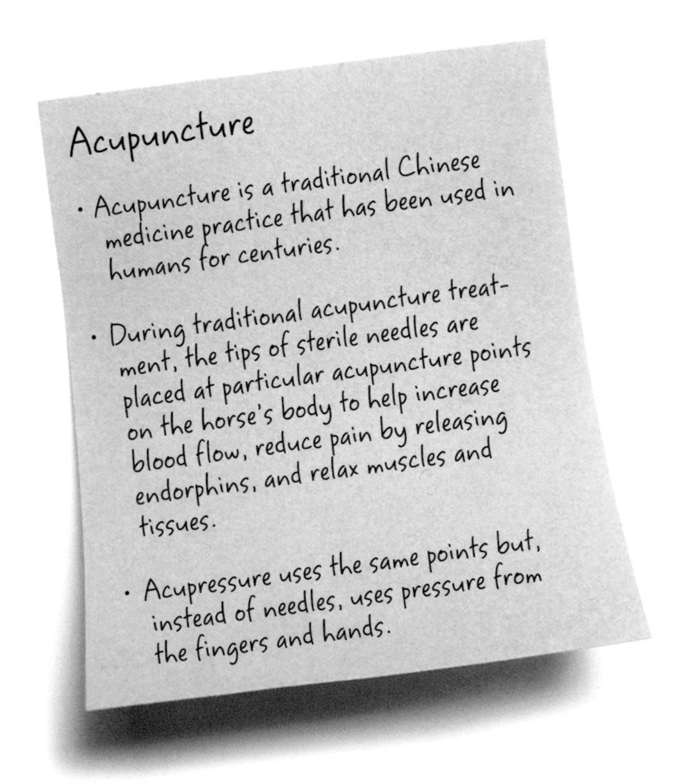

Chiropractic and acupuncture are the most common complementary therapies along with massage (which owners often learn to perform themselves). Finding a trained and reputable chiropractor or acupuncturist can be the hardest part. While many people may call themselves an equine chiropractor or acupuncturist, it's safest to hire a veterinarian who is also trained in the complementary therapy you're seeking. Your regular vet or a fellow horse owner may be able to provide a recommendation, or you can contact the International Veterinary Acupuncture Society and the American Veterinary Chiropractic Association. Once you find a veterinarian practitioner in your area, don't be shy about asking for additional personal recommendations.

Chiropractic adjustments in horses are similar to chiropractic treatments in humans except, of course, the practitioner comes to your stable and adjusts the horse standing up! Keep in mind that back pain can cause lameness, and lameness can cause back pain. Chiropractic treatment may help reduce pain, improve flexibility, and help the horse travel more evenly.

Acupressure

- Acupressure is similar to acupuncture, but pressure from fingers and hands is used to stimulate the points rather than needles.
- Unlike the other complementary therapies discussed here, horse owners willing to do a little bit of research can use acupressure or massage on their own without the assistance of a professional.
- There are sources devoted to do-it-yourself equine acupressure and massage.
- Acupressure can be used alone or combined with massage techniques.

Homeopathy

- Homeopathy is geared toward stimulating the body's defense mechanisms by administering minute doses of natural substances that in higher doses would cause the symptoms currently affecting the patient.
- "Remedies" can be specially prepared taking into account the individual animal and his circumstances, or there are general remedies for certain common conditions.
- Homeopathy is different from herbology, which is based on the medicinal use of herbs.

CREATING A FIRST-AID KIT

When your horse injures himself, there's no time to drive around town gathering supplies, so be prepared

Every horse owner should have a well-stocked first-aid kit. Most items can be bought from the drug store, tack store, or ordered from a catalog or Internet site. However, certain medications you'll need to get from your veterinarian.

Start off your first-aid kit with a large plastic container with an airtight lid. This will serve as a storage container for your first-aid kit and can also be used for mixing your wound cleaning solutions. Before you fill the container, use a permanent marker to mark on the side how full the container will be with one quart of water. Then make several baggies, each with a half tablespoon of salt that can be added to the one-quart of water for cleaning wounds or irrigating an eye.

Bandaging Materials

- Your first-aid kit should also include these bandaging materials:
  - Sterile, nonstick dressing, roll gauze, and bandage scissors.
  - Disposable cotton batting for use as the padding material and self-adhesive, stretchable veterinary wrap for holding the bandaging in place.
- Also remember to replace used items, keep track of expiration dates on medications, and keep your first-aid kit fully stocked.

Other Essentials

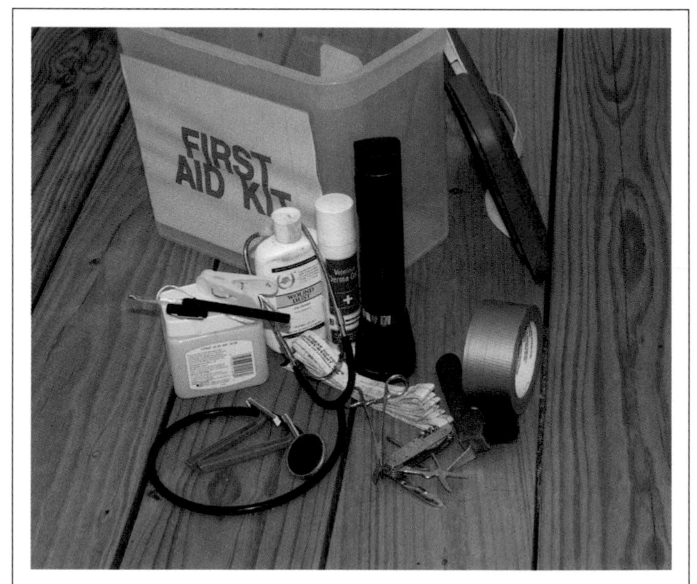

- Your first-aid kit should also include:
  - A rectal thermometer (the 5-inch large-animal type or human digital thermometer) and a stethoscope to count heart rate and to check intestinal sounds.
  - A multipurpose tool, a hoof pick, and tweezers or forceps.
  - Duct tape, a hoof boot, fly repellent packets, and a flashlight with fresh batteries.

Next, add your wound cleaning materials to the kit. You'll need disposable razors or scissors for hair removal and disposable gauze sponges for scrubbing. Also include antiseptic (povidone iodine or chlorhexidine) solution, which can be added to the saline solution for rinsing and irrigating wounds. Use 10ml povidone iodine or 20ml chlorhexidine per liter (or quart) of salt water. It's also a good idea to include antiseptic (povidone iodine or chlorhexidine) surgical scrub for scrubbing wounds. Lastly, you'll need a 35cc or 60cc syringe for wound or eye irrigation and some disposable gloves.

For treating wounds after they've been cleaned, your kit should include a topical, water-soluble antibiotic ointment. A diaper rash ointment (Desitin) can be used for saddle sores, rope burns, or scratches. Clean padding and wrap for a standing bandage (pages 188 & 189) can be used to appply a tight pressure bandage over the wound to control bleeding.

Instant hot and cold packs are also a good idea to include in any first-aid kit, and additional essentials are discussed below.

Prescription Medications

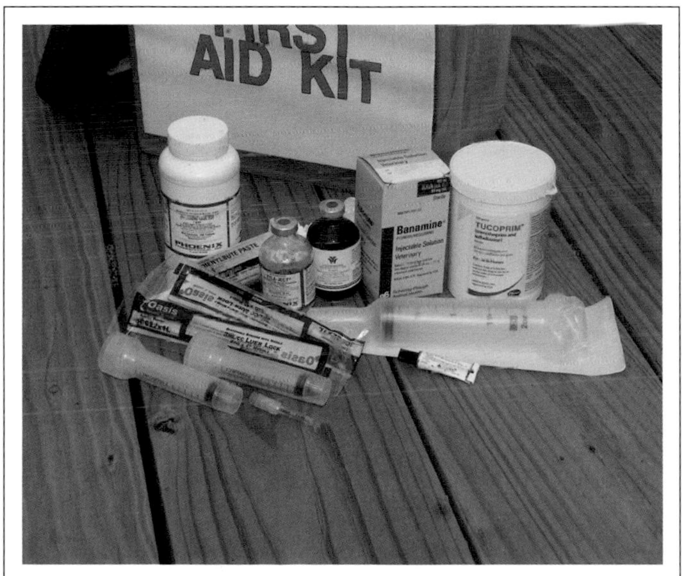

- You'll need to obtain the following perscription medications from your veterinarian along with directions for use:
  - Nonsteroidal antibiotic eye ointment, broad-spectrum oral antibiotics, and oral dose syringe.
- Nonsteroidal anti-inflammatory systemic medication (phenylbutazone, flunixin meglumine, or ketoprofen) for pain and swelling, short-acting intramuscular sedative for pain relief, and epinephrine to counteract allergic reaction.

First-Aid Kits

- There are ready-made first-aid kits you can purchase; just be sure to check the contents.
- It's a good idea to take your kit with you when you trailer your horse to a show or out on a campout, or have a separate kit in your trailer.
- Ready-made kits can be purchased especially for the trailer and smaller kits for trail riders, or you can make your own.

VITAL SIGNS

It's important to know how to check and monitor your horse's vital signs

While not all hoof and leg problems affect a horse's vital signs, many painful conditions and issues, such as infection and tying up, can alter one or more of your horse's vital signs. Anytime you suspect a problem, take your horse's vital signs and write them down—along with any other pertinent information—to convey them to the vet.

To find a heart rate, using a stethoscope (inexpensive ones are available through catalogs or at a pharmacy), count the number of beats per minute (bpm) by placing the end of the stethoscope on the body wall just behind and at the level of the left elbow. Each lub-dub you hear counts as one beat. The normal resting heart rate for a horse can range from twenty-

Rectal Temperature

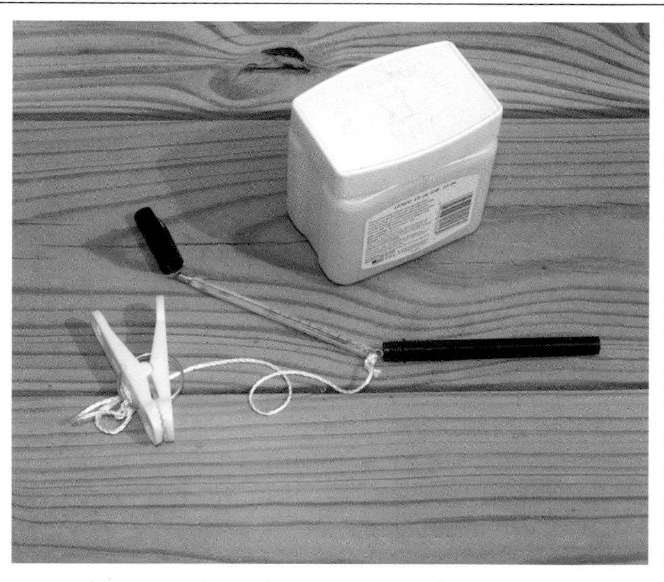

- Attach a string with a clip to hook onto the horse's tail so you don't have to hold the thermometer with your fingers.
- Shake down nondigital thermometers to less than 96 degrees Fahrenheit. Then, lubricate the thermometer with petroleum jelly.
- Leave the thermometer in place for about two minutes.
- Normal temperature for an adult horse is 99.5–101.5 degrees Fahrenheit. Any reading above this in a resting horse (twenty minutes after exercise) is considered a fever.

Mucous Membranes

- Lift the horse's upper lip and look at the gum color above the teeth. Membranes should be pink and moist, like the color you see beneath your fingernails.
- Pale pink indicates decreased circulation, anemia, blood loss, or systemic illness.
- Bright red or muddy-looking membranes can indicate shock.
- Yellow-hued membranes are often associated with liver stagnation or may simply indicate the horse has been eating legume plants or not eating at all.

eight to forty bpm. Pain will often drive a horse's heart rate up to sixty-four bpm. Critical problems like shock or an intestinal twist will cause a horse's heart rate to rise over eighty bpm, and it will remain elevated. Sometimes a spasm of pain will cause transient elevation, so check the heart rate again in ten to fifteen minutes. Rapid breathing, obvious signs of pain and distress, or depression often accompany a rapid heart rate.

To find the respiratory rate, count the number of respirations per minute. Do this by counting each diaphragm lift visible in the flank area. Or, hold your hand in front of the horse's nostrils and count each breath. The number of respirations per minute gives an indication of whether the horse is normal or stressed. Normal respiratory rate in a resting horse is twelve to twenty-four breaths per minute. Arduous exercise, or overheating from fever or heat stress, will cause an elevation in the respiratory rate. High ambient temperatures, particularly with high humidity, may elicit a rapid respiratory rate in the absence of any problem.

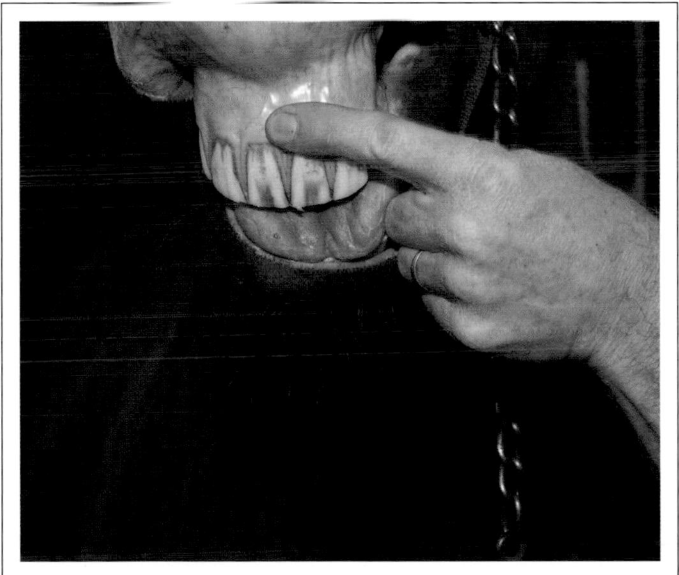

Capillary Refill Time

- Blanch the membranes of your horse's upper gum by pressing it with the tip of your finger and note how quickly the pink color returns.
- Normal capillary refill time (CRT) should return the gums to a pink color in less than two seconds.
- Delayed CRT indicates cardiovascular compromise, often associated with dehydration or acute blood loss.

Hydration

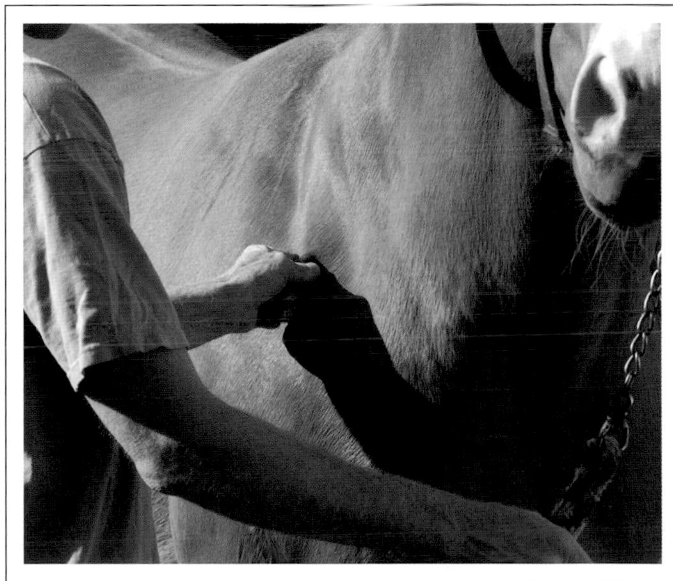

- To determine a rough estimate of a horse's hydration, grab a fold of skin near the point of the shoulder and note how quickly it snaps back into position. Normally, the skin snaps back immediately.
- Skin that stays "tented" for a moment or returns to normal slowly indicates dehydration.
- Skin that remains tented and refuses to return to its normal position may represent serious, life-threatening dehydration.

ALLERGIES, BITES, AND STINGS

Allergies, bites, and stings often require first-aid treatment and careful monitoring

A horse's legs, like the rest of his body, are susceptible to bites, stings, and allergies. Horses can develop allergies just like people do, with the problem coming on all of a sudden or developing slowly. Hives are one possible sign and are often associated with more progressive forms of allergy. Hives are localized, raised bumps visible on the skin, and can vary in size and location. Other possible signs of allergies that can occur alone or in combination include depression, mild fever, swelling near the throatlatch, swelling in the limbs (or abdomen, sheath, or udder), difficulty breathing, and anaphylaxis (extreme restlessness with sweating and potential for death).

Ticks

- Ticks can carry Lyme disease and should be removed from horses promptly.
- To remove a tick, wear gloves and grasp the tick behind its head with your fingers or blunt-tipped forceps.
- Pull gently, being careful not to leave the tick's head embedded in the horse's skin.
- Kill the tick by burning him, putting him in alcohol, or flushing him down the toilet, then properly treat the bite wound.

Wasps and Bees

- Multiple bites from stinging bees, wasps, hornets, or yellowjackets can cause adverse reactions in horses.
- Avoid known bee hives, and don't antagonize these insects by swatting at them. Instead, calmly leave the area.
- Topical treatments for stings can include applying calamine lotion or a mixture of baking soda and water.
- If your horse has been stung multiple times, he may need veterinary attention, including systemic anti-inflammatory medication and antihistamines.

Like people, horses can develop allergies to just about anything, including inhaled pollens or molds, insect bites or stings, medications, fly sprays, certain plants, feed supplements (especially high-protein sources), pine shavings, soap used on blankets and tack, and antiseptic scrubs.

If your horse suddenly develops an allergy, remove all food and supplements (especially anything new) from his diet other than grass hay. Also temporarily eliminate medications and fly repellents. If riding or sweating seems to make the horse worse, take a break from activity. Wash all tack and equipment, rinsing them free of soap. If the horse's condition doesn't steadily resolve within twelve to twenty-four hours or worsens, contact your veterinarian for further instructions or treatment. Severe cases of allergic reaction can trigger anaphylaxis and cause the airways to swell shut. These cases require immediate veterinary attention, and the horse will likely need to be quickly treated with epinephrine and corticosteroids as a life-saving measure.

Bites and stings are discussed below. See page 185 for more information on fly and insect sensitivity.

Spider Bites

- Brown recluse spiders and black widows pose the most danger and are often found in woodpiles, barns, and sheds.
- A brown recluse bite can set off a localized reaction; a bite from a black widow can elicit both a localized and systemic reaction.
- Signs of a bite include inflammation in the area, itchiness, dead skin, and sensitivity.
- Your vet can prescribe anti-inflammatory medications and antihistamines and remove affected tissue so that it doesn't spread.

Snakebites

- Rattlesnakes are a problem throughout the country, and a grazing horse may surprise one and get bit.
- While many bites are dry (no venom), the bite can introduce bacteria, leading to infection.
- Bites are often associated with rapid swelling. A bite near the nose can obstruct the airways.
- If you suspect a snakebite on any part of your horse's body, seek veterinary help immediately.

LOOSE AND LOST SHOES

If your horse is shod, it's important to know how to deal with loose, bent, or lost horseshoes

Loose, bent, or lost horseshoes require first-aid until the farrier can return to fix the problem. (Always put a call in to your farrier immediately.) Don't ride a horse with a loose, bent, or missing horseshoe. A loose or bent shoe can put the horse in danger of stepping on one of the shoe nails. If you're out on the trail, use the methods described here, and then lead your horse home. These methods can work using a multipurpose tool and a rock if you're out on trail or using common items you'd have in your toolbox at home.

A sprung, twisted, or bent horseshoe will need to be removed. To remove a shoe, you'll first have to unclench the nails. One method is to file the clinches away until the ends

Tightening a Loose Shoe

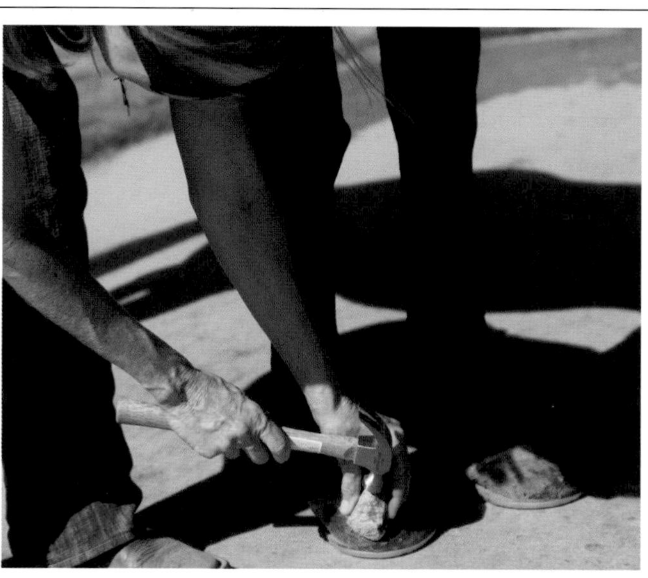

- If a shoe is loose, you can try firming up the clinches.
- Place a rock or other hard object directly beneath the head of one of the loosened nails, then take a screwdriver (or the screwdriver in the multipurpose tool) and pound it with a round rock or hammer directly over the nail clinch, which should do the trick.
- A properly fitted hoof boot placed over the shoe can keep the horse from losing it altogether.

Emergency Hoof Protection

- If your horse loses a shoe or has one removed on trail, and you don't have a hoof boot, you can make a protective barrier for your horse's hoof by wrapping the foot in soft material (like a diaper or T-shirt) and then covering it with Vetrap or duct tape.
- Then lead your horse slowly home, trying to walk on soft ground whenever possible and reapplying the tape as needed.

of the nails are flush with the hoof wall. Use your multipurpose tool's file if you're out on trail or a rasp if you have one at home. Be sure not to file or rasp a hole in the hoof wall. Another method is to unclench the nails by driving the end of a flat-bladed screwdriver or chisel under the clinch to straight it. Once the clinches have been removed or straightened, pry the shoe off a little at a time with pliers, banging the shoe back down with a rock or other hard object to expose the top of each horseshoe nail. Then pluck the loosened nails out using the pliers. Be sure no nails are broken off inside the hoof wall. If one is, try to remove it.

If you trail ride, it's a good idea to always carry a hoof boot sized for an unshod front hoof in case you have to remove a shoe or loose one out on trail. It's also a good idea to have such a boot in your first-aid kit at home.

Farrier Tools

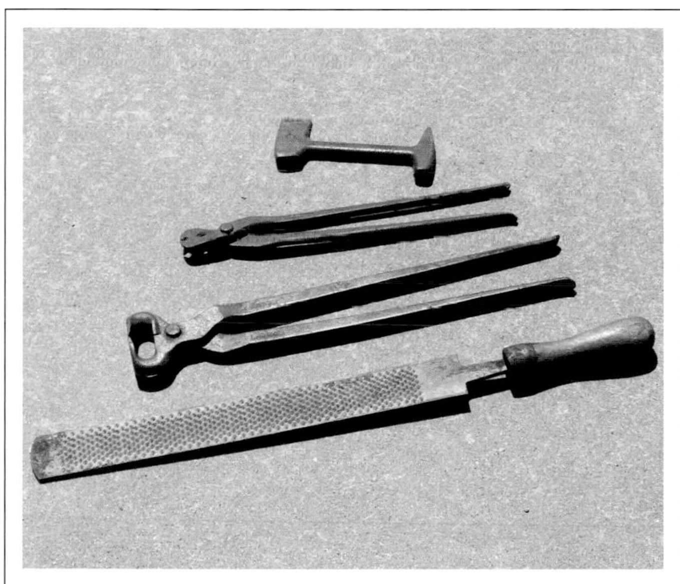

- Some horse owners choose to invest in a basic set of farrier tools to keep on hand for emergency shoe problems.
- Tools can include a rasp, pull-offs/shoe-pulling pliers, or crease nail-pullers (which remove the nails to free the shoe rather than pulling the shoe off).
- If you purchase these tools, ask your farrier for a demonstration on how to use them in case you must remove a shoe that becomes twisted or bent.

Prevention

- Properly maintaining a horse's hooves, including regular visits from the farrier will help prevent lost, loose, or bent shoes.
- If your horse's hooves are overdue for a shoeing (shown here) or not looking healthy, wait to ride until they've been attended to.
- Shoes that are lost in the corral, pasture, or arena need to be found. Stepping on nails or shoe clips can injure a horse.

HOOF WRAPS

Treating hoof injuries may require the application of a hoof wrap

There are a number of occasions that may require covering the hoof to protect it and keep it clean while it's being treated. Abscesses and puncture wounds are two examples of issues that may require this treatment. The vet may recommend a hoof boot. However, hoof boots come in different sizes, and if you have multiple horses, you may not have one that fits every horse. If an appropriate hoof boot is not available, the vet may suggest wrapping the hoof as an alternative. Although wraps can be bulky, some form of wrap may also be used in combination with a hoof boot (again depending on the size of boot available).

Because of their shape and the fact hooves carry a great deal of weight and suffer wear and tear, it can be difficult to apply a wrap correctly and have it stay on; therefore, it's helpful to practice applying a hoof wrap before you actually need to use the skill. (See below for step-by-step instructions.)

Hoof Bandage 1

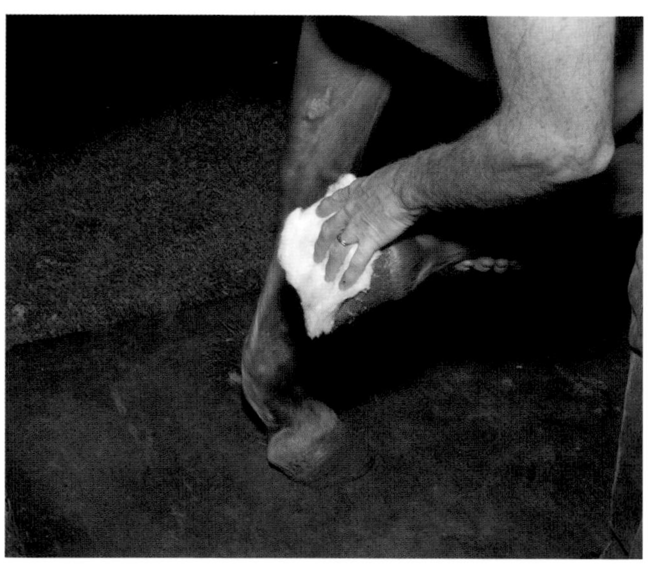

- Before applying your hoof bandage, you may first apply a topical product per your veterinarian or farrier's instructions, such as a non-stick dressing or povidone-soaked piece of cotton over the area of the hoof that is of concern.
- Then, while holding the hoof up, lay a piece of disposable cotton sheet batting over the bottom of the hoof so that it sticks up around the sides and behind the heel bulbs.

Hoof Bandage 2

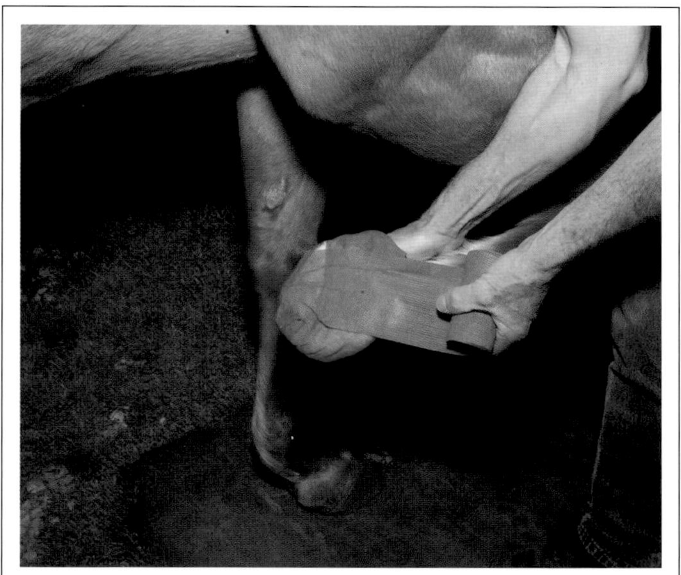

- Once the padding is in place, use self-adhesive bandaging material such as Vetrap and tape over the bottom of the hoof.
- Layer the bandaging material over the bottom of the hoof and around the edges.
- Use a figure-eight pattern across the back of the heel bulbs to keep the bandage from slipping.

Premade, easy fasten hoof wraps, hoof soakers, and a variety of hoof boots are available commercially. Most hoof boots are designed for exact fit, which requires careful measurement. Hoof boots should generally be purchased to fit a barefoot front hoof. This way, if a shod horse loses a shoe or is being treated for a hoof problem, the boot will fit him correctly.

Horses with abscesses or puncture wounds may need to have their hooves soaked in warm water and Epson salts. Some people create their own hoof soakers or hoof poultice wraps using baby diapers (plastic or the kind designed for swimming pools). After applying the diaper around the hoof as you would apply it to a baby's bottom, a few swipes of duct tape are used to shore up the sides and reinforce the bottom. Just be sure to consult your veterinarian and farrier, and follow their treatment guidelines carefully.

Hoof Bandage 3

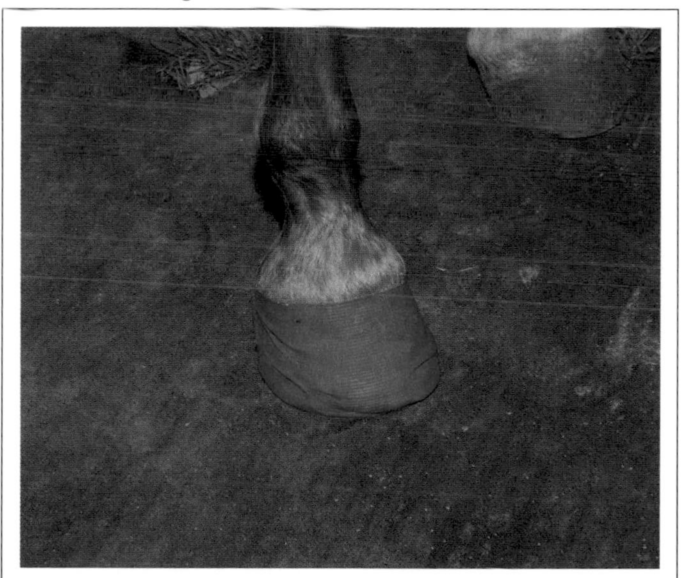

- It's best to keep the bandaging material below the coronary band in the front of the hoof.
- If the bandage does cover the coronary band, be sure not to apply it too tightly.
- As with other bandages, it's best to keep your horse confined to a clean stall and check the bandage regularly.
- Your farrier or veterinarian can advise you on how long the bandage will be used and how often to change it depending on the problem you're treating.

Hoof Bandage 4

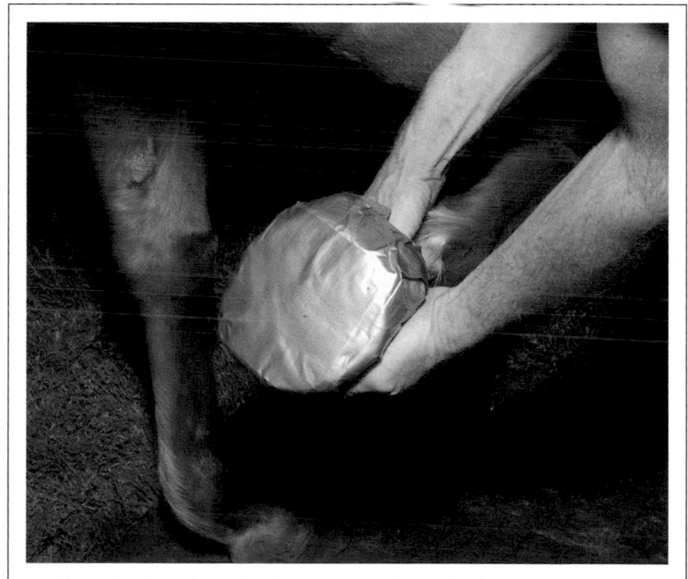

- After the bandage has been secured using self-adhesive Vetrap, apply duct tape across the bottom of the hoof.
- Duct tape helps minimize wear on the bandage.
- Reapply duct tape as needed to protect the bottom of the wrap.
- When it's time to remove the hoof bandage, use a pair of blunt-end bandage scissors.

HORSE ANATOMY

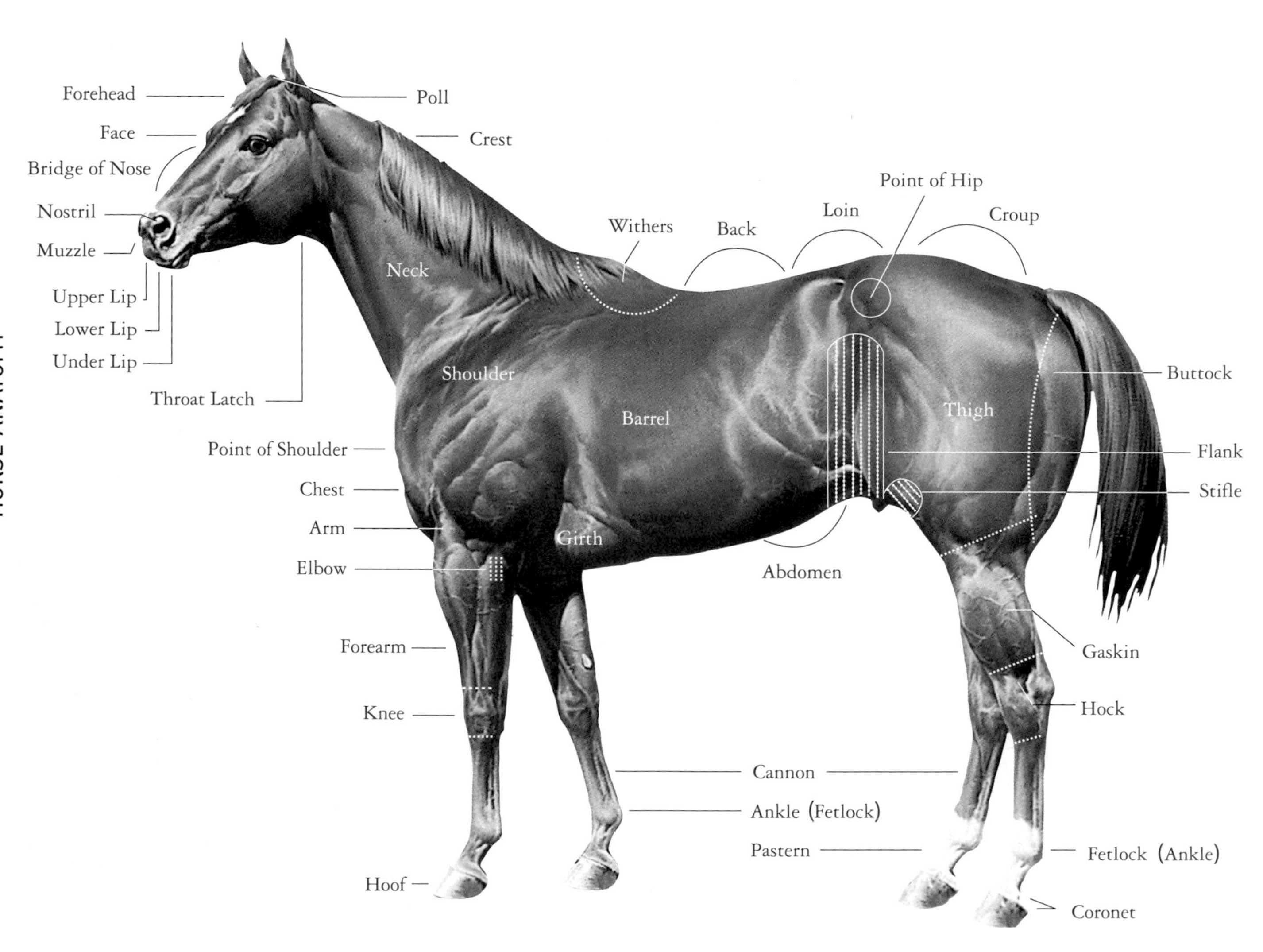

Skeleton

1. Premaxilla
2. Maxilla
3. Nasal
4. Frontal
5. Parietal
6. Occipital
7. Mandible
8. Cervical vertebrae
9. Cartilage of scapula
10. Scapular spine
11. Scapula
12. Humerus
13. Olecranon
14. Ulna
15. Radius
16. Carpus
17. Metacarpals
18. Phalanges of forefoot
19. Costal cartilages
20. Ribs (18)
21. Thoracic vertebrae
22. Lumbar vertebrae
23. Tuber sacrale
24. Ilium
25. Ischium
26. Sacrum
27. Coccygeal vertebrae
28. Femur
29. Patella
30. Fibula
31. Tibia
32. Tuber Calcis
33. Tarsus
34. Metatarsus
35. Phalanges of hindfoot

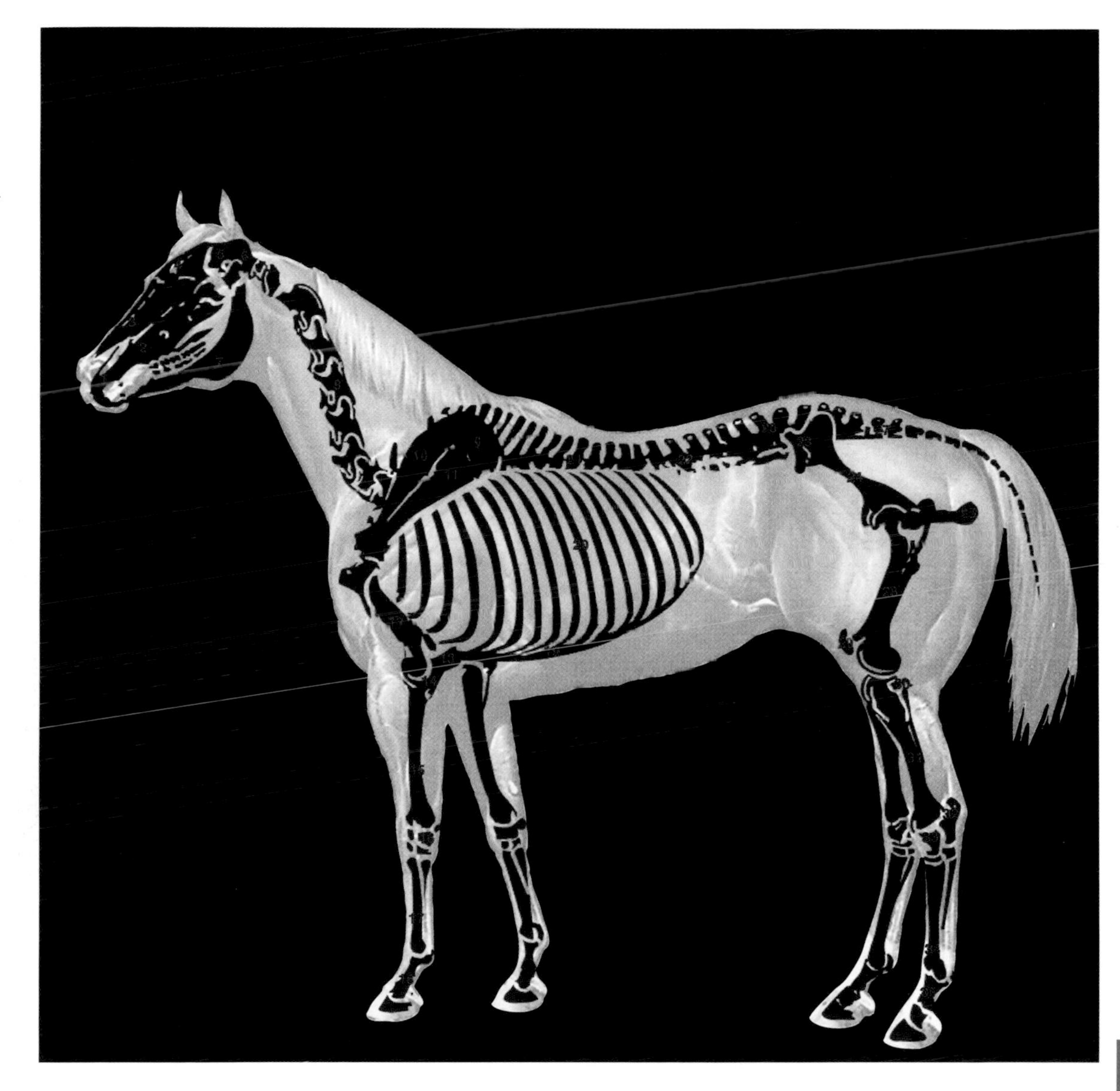

Muscles

1. Levator nasolabialis
2. Zygomaticus
3. Buccinator
4. Facial vein
5. Levator labii sup proprius
6. Masseter
7. Scutularis
8. Parotido-auricularis
9. Rhomboideus
10. Jugular vein
11. Splenius
12. Sterno-cephalicus
13. Brachiocephalicus
14. Serratus cervicis
15. Trapezius
16. Pectorals, deep
17. Supraspinatus
18. Deltoid
19. Pectorals, superficial
20. Biceps brachii
21. Brachialis
22. Ext. carpi radialis
23. Common digital ext.
24. Deep flexor
25. Ulnaris lateralis
26. Serratus thoracis
27. Triceps brachii
28. Latissimus dorsi
29. Obl. abdominis ext.
30. Aponeurosis of obl. abd .ext.
31. Lumbodorsal fascia
32. Gluteal fascia
33. Tensor fascia latae
34. Gluteal fascia
35. Fascia lata
36. Biceps femoris
37. Semitendinosus
38. Long digital extensor
39. Soleus
40. Lat. digital extensor
41. Gastrocnemius
42. Saccrococcygeus

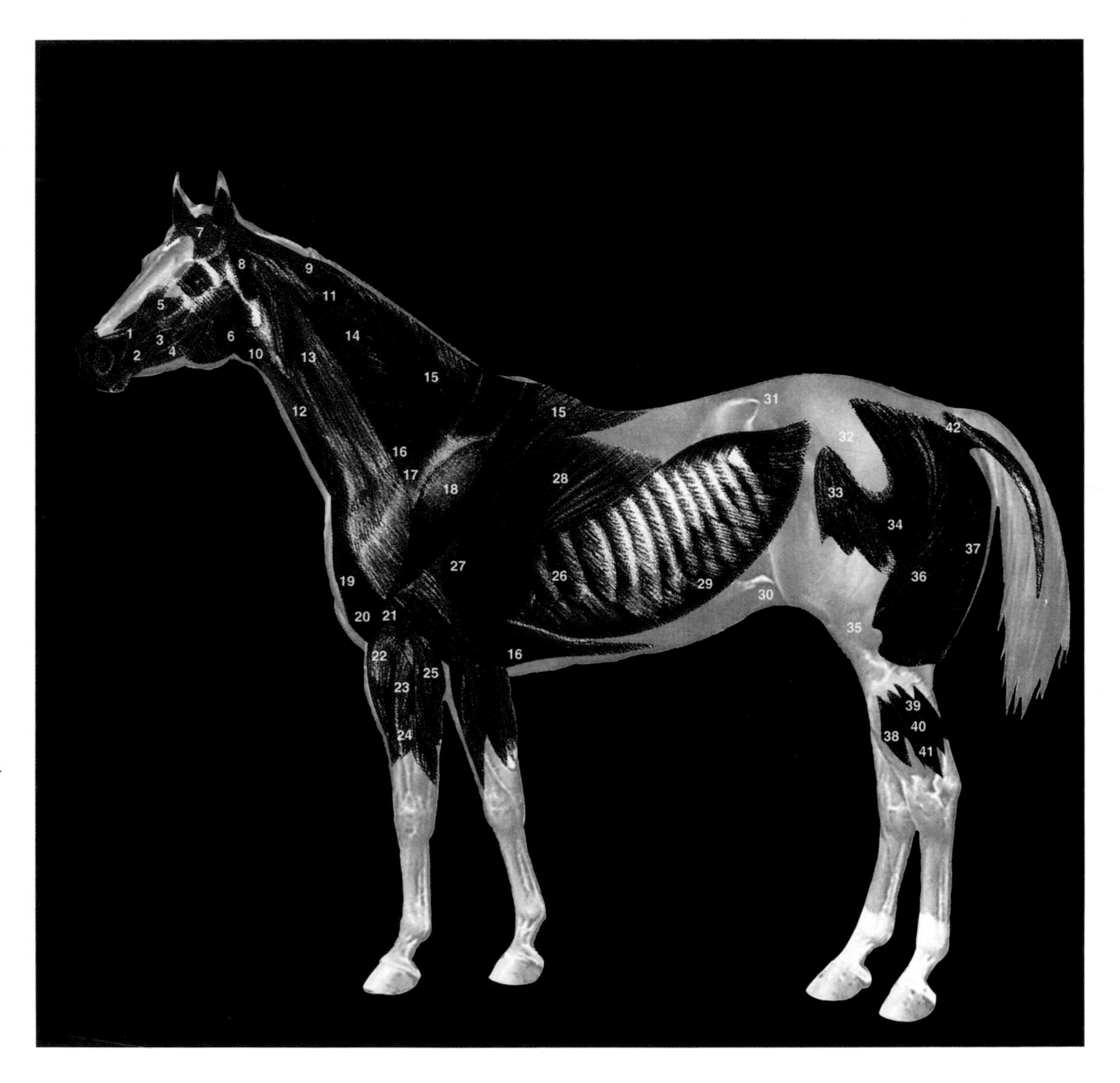

1. Buffs
2. Central sulcus of frog (spine of frog) (frog stay)
3. Angle of wall
4. Bars
5. Collateral sulcus
6. White Line
7. Apex of frog
8. Wall
9. Sole

Normal Forefoot

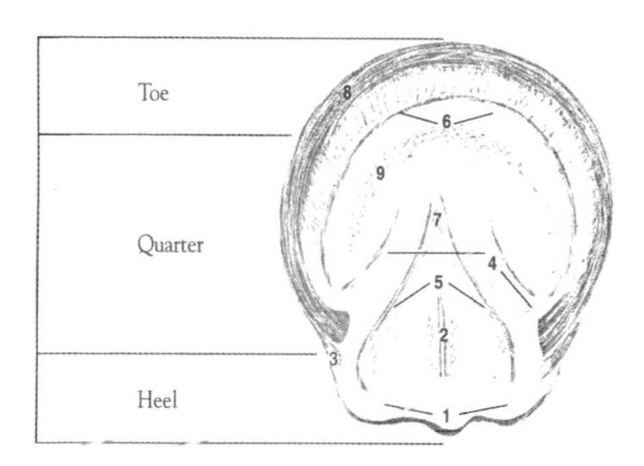

Normal Hindfoot

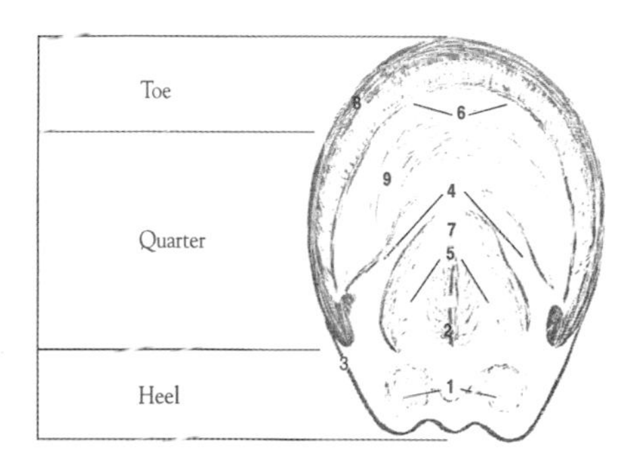

Lateral View of Lower Leg Bones

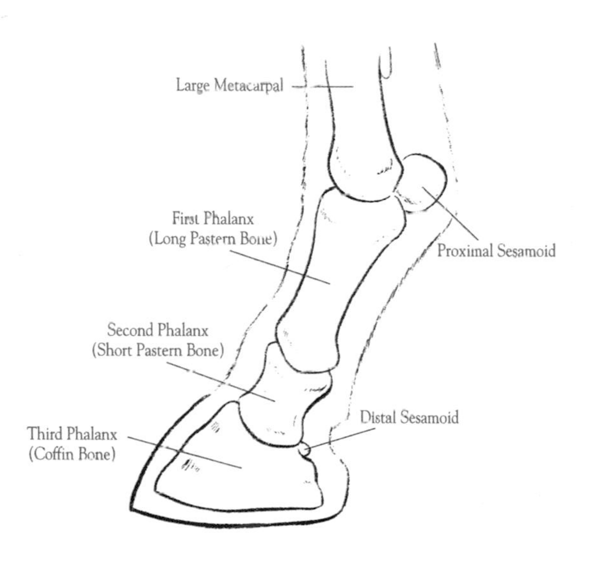

RESOURCES

When it comes to learning about the leg and hoof of the horse, there are a lot of additional resources available online. These Web sites are just a sampling. Many of these sites offer informative articles that can help you in your research on hoof and leg care, and lameness issues. Please note, however, that the author and publisher do not specifically endorse every site listed and cannot guarantee the accuracy of the information found throughout each site.

General Information

Equine Research, *Lameness: Recognizing and Treating the Horse's Most Common Ailment,* The Lyons Press, 2005

Ross, Michael W., and Sue J. Dyson, *Diagnosis and Management of Lameness in the Horse,* Saunders, 2002

Stashak, Ted S., and Robert Adams, *Adams' Lameness in Horses,* Wiley-Blackwell, 2002

Horse Care and Health Web Sites

Acreage Equine
www.acreageequines.com

Arizona Horse pages
www.horses-arizona.com

Ask the Farrier
www.askthefarrier.com

Bear Creek Mobile Veterinary Service Informational Brochures page
www.bearcreekvet.net/client_education.html

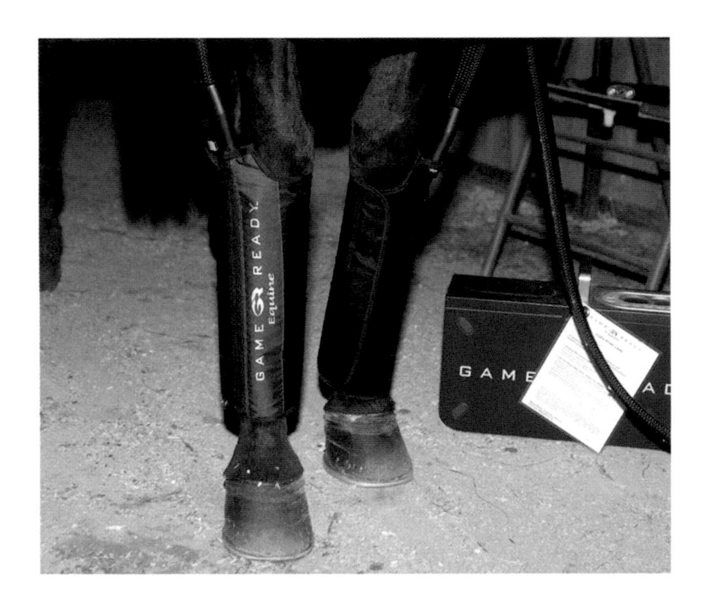

Bloodhorse.com
http://health.bloodhorse.com

Cherry Hill's Horsekeeping® LLC Books and Videos
www.horsekeeping.com

Davie County Large Animal Hospital, PA (click on "Brochures")
www.dclahdvm.com

Douglas Novick, DVM
www.novickdvm.com

Dr. L. Clarke Cushing Articles page
www.lclarkecushingvmd.com

Equine Natural Therapy
www.equinenaturaltherapy.com

EquineSite
www.equinesite.com

Equisearch
www.equisearch.com

Equiworld
www.equiworld.net

Equusite
www.equusite.com

Extension horses page (articles, plus find your local office, etc.)
www.extension.org/horses

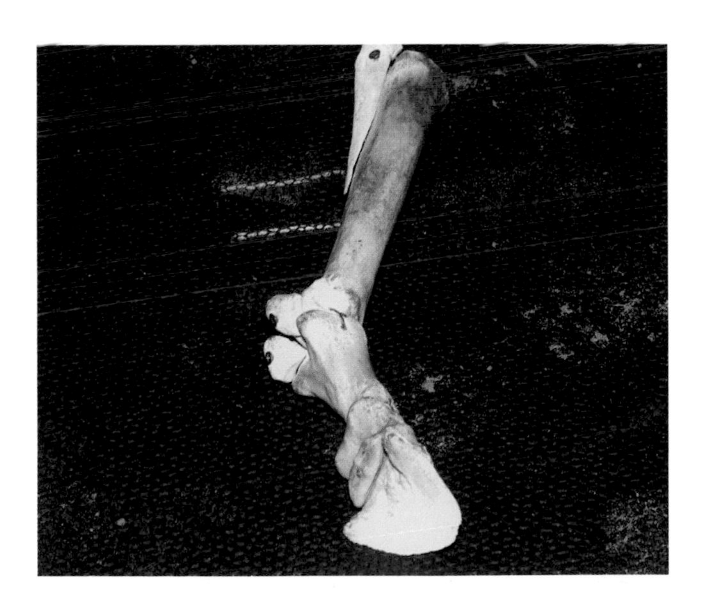

Hoofcare and Lameness
www.hoofcare.com

Horse and Hound
www.horseandhound.co.uk

Horse Channel
www.horsechannel.com

HorseCity.com
www.horsecity.com

Horses and Horse Information
www.horses-and-horse-information.com

Horse Previews Magazine
www.horse-previews.com

HorseQuest.com
www.horsequest.com

Laminitis page
www.laminitis.com

Michigan State University College of Veterinary Medicine
Animal Health Info page
http://old.cvm.msu.edu/hinfo.htm

MyHorse.com
www.myhorse.com

New Rider
www.newrider.com

Northern Virginia Equine Podiatry
www.equipodiatry.com

Pacific Crest Equine (click on "Equine Health")
www.pacificcrestequine.com

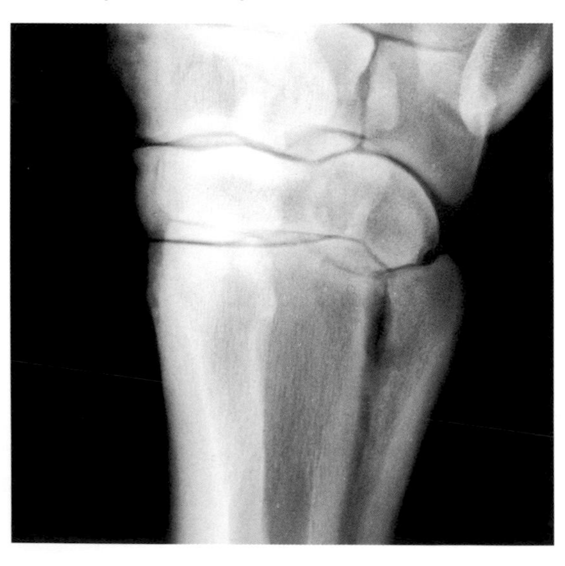

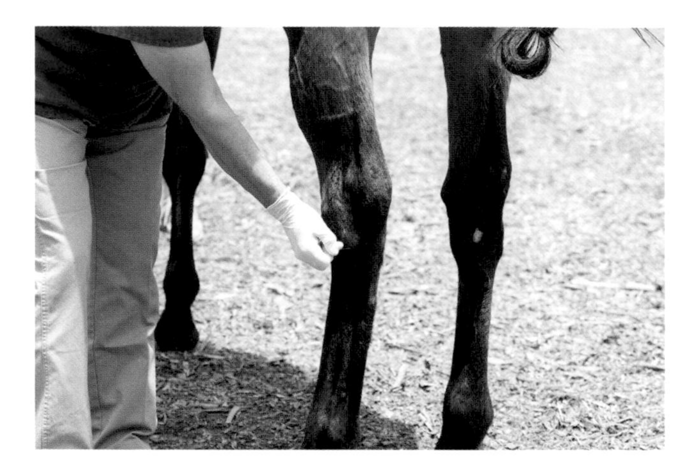

Penn State College of Agricultural Sciences Publications
http://pubs.cas.psu.edu

Purdue University Equine Health Update Newsletter
http://vet.purdue.edu/esmc/articles.html

Rural Heritage (Vet Clinic section)
www.ruralheritage.com

SaferGrass.org - Education for the Prevention
and Treatment of Laminitis in Horses
www.safergrass.org

Sylvia Greenman, DVM
www.greenmanequine.com

The Atlanta Equine Clinic Veterinary Medicine News page
www.atlantaequine.com/pages/vet_news.html

The Complete Rider
www.completerider.com

The Equine Center (click on "Client Education")
www.theequinecenter.com

The Horse
www.thehorse.com

The Horse Magazine (Australia)
www.horsemagazine.com

The Liphook Equine Hospital Information page
www.liphookequinehosp.co.uk/infosheets.htm

The Merck Veterinary Manual
www.merckvetmanual.com

University of California, Davis Center for Equine Health
www.vetmed.ucdavis.edu/ceh

University of Illinois at Urbana-Champaign College of Veterinary Medicine Pet Columns
www.cvm.uiuc.edu/petcolumns

VeterinaryPartner
www.veterinarypartner.com

VeterinaryWatch
www.veterinarywatch.com

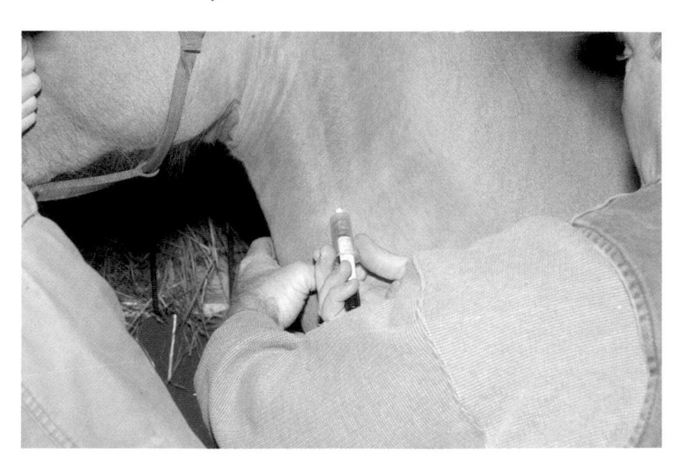

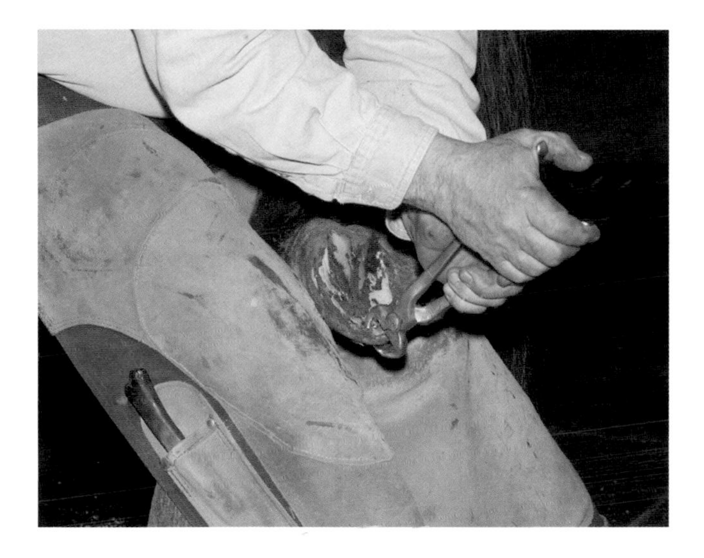

Hoof Trimming and Shoeing Web Sites

American Association of Farriers
www.americanfarriers.org

American Farriers Journal
www.lesspub.com/afj

American Hoof Association
www.americanhoofassociation.org

Anvil Magazine
www.anvilmag.com

Ask the Farrier
www.askthefarrier.com

Association for the Advancement of Natural Horse Care Practices
www.aanhcp.net

Brotherhood of Working Farriers Association
www.bwfa.net

Hoofcare & Lameness
www.hoofcare.com

Hoof Rehabilitation Specialists Ivy and Pete Ramey
www.hoofrehab.com

Lady Farrier
www.ladyfarrier.com

Nanric Inc.
www.nanric.com

Penzance Progressive Natural Hoofcare
www.barefoottrim.com

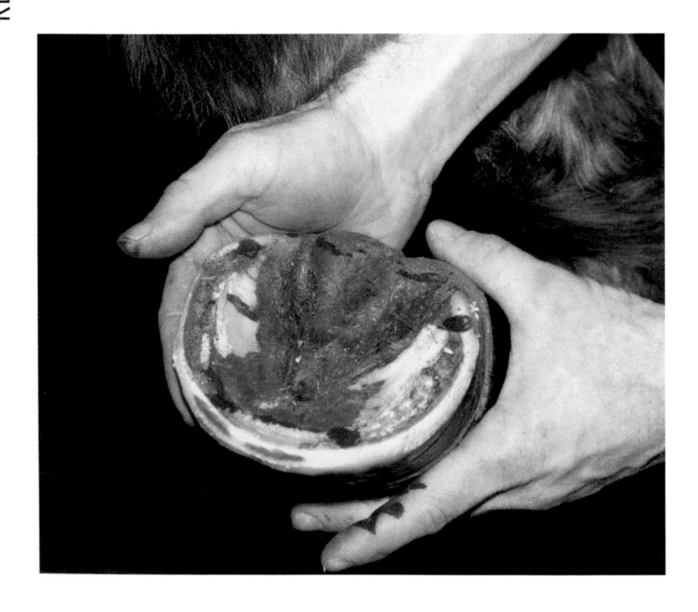

Strasser Hoofcare North America/Equine Soundness Association
www.strasserhoofcare.org

The Farrier and Hoofcare Resource Center
www.horseshoes.com

The Home of Natural Balance Hoof Care
www.hopeforsoundness.com

The Horse's Hoof
www.thehorseshoof.com

World Wide Farrier Directory
www.farriers.com

Veterinary Association Web Sites

The American Association of Equine Practitioners
www.aaep.org

American College of Veterinary Surgeons
www.acvs.org

American Veterinary Chiropractic Association
www.avcadoctors.com

The American Veterinary Medical Association
www.avma.org

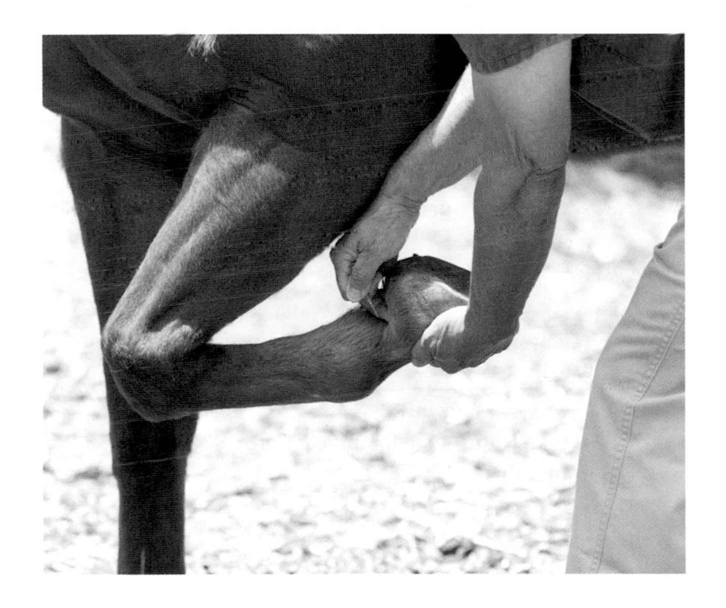

Conformation Booklets

University of Arkansas Cooperative Extension Service Horse Conformation Analysis booklet
www.uaex.edu/Other_Areas/publications/PDF/FSA-3029.pdf

Washington State University Extension Horse Conformation Analysis booklet
http://cru.cahe.wsu.edu/CEPublications/eb1613/eb1613.pdf

GLOSSARY

Physical/directional terms to know in reference to legs and hooves

Please note that these definitions are not general definitions but define the use of these terms in reference to legs and hooves in particular.

Caudal and Posterior
Refer to the back part of the leg or hoof.

Cranial and Anterior
Refer to the front part of the leg or hoof.

Distal
Refers to being farther from the body—think lower and closer to the ground.

Lateral
Refers to the outsides of the legs and hooves—the sides that face out.

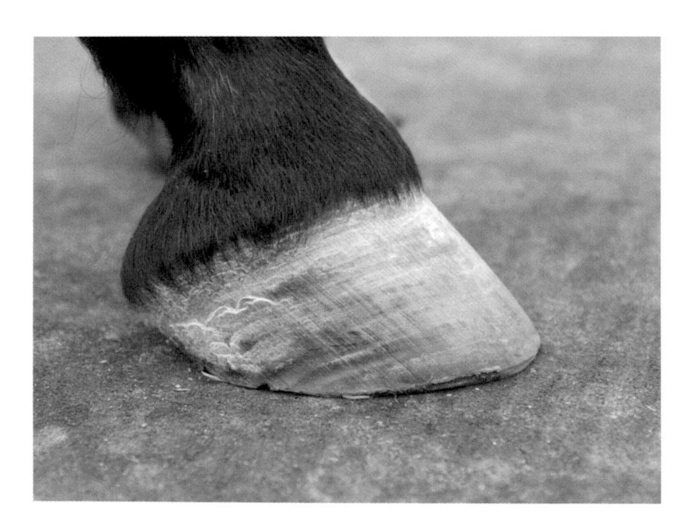

Medial
Refers to the inside of the leg and hoof—the side that faces the opposite leg and hoof.

Palmar
Refers to the backside and underside of the forelimb.

Plantar
Refers to the back and underside of the hindlimb.

Proximal
Refers to being closer to the body, which in terms of legs and hooves means higher up.

Quarters
The quarters are the sides of the hooves. The toe is in front, the heel in back, and the quarters are the sides.

Ventral
This term refers to the underside/bottom side.

Fracture Terminology

Articular
These fractures involve a joint surface and are usually more severe than nonarticular (not involving a joint surface).

Comminuted
Means the fracture is fragmented, which is more severe than a "simple" fracture.

Complete
These fractures go all the way through the bone and are more severe than incomplete or stress fractures.

Compound
These fractures are open, meaning the skin has been broken, and the bone may be sticking out.

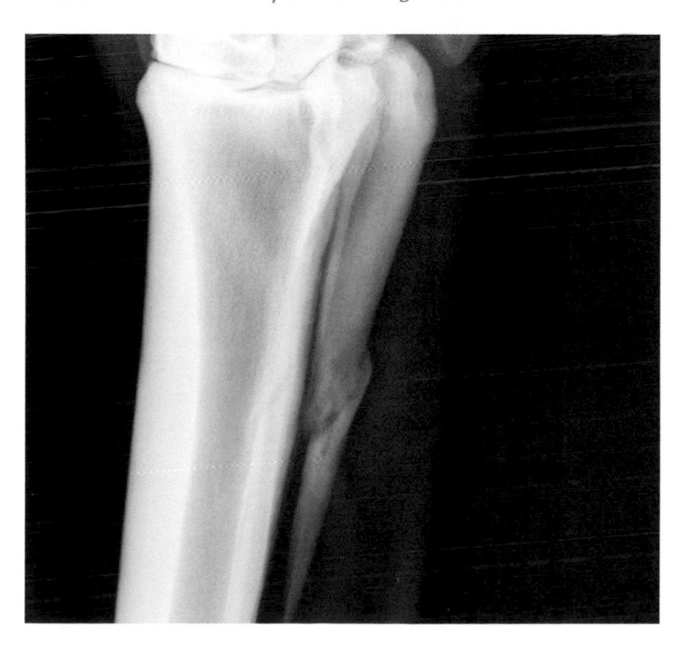

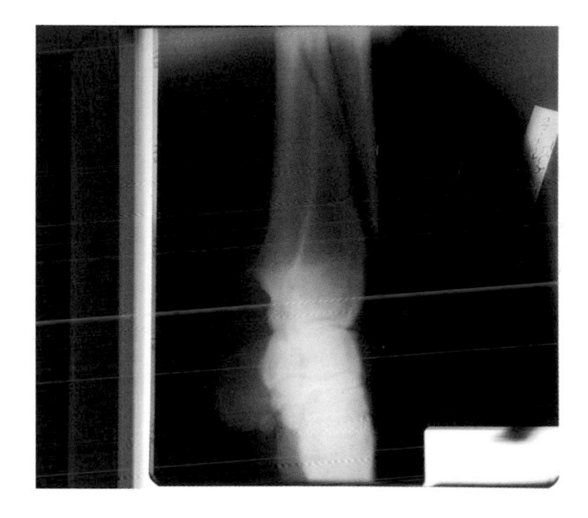

Displaced
A fracture that's displaced means the bone is knocked out of place and is more severe than nondisplaced (where the bone is fractured, but the two parts remain in place).

Sagittal
This term refers to when the fracture runs vertically.

Slab
You will usually hear the term "slab fracture" in regard to bones in the tarsus (hock) or carpus (knee). Slab fractures usually involve the full thickness of the bone and two articular (joint) surfaces. Slab fractures are often treated using a screw to reattach the slab to the parent bone when possible.

Stress
These fractures are also sometimes called "hairline" because they create very minute cracks in the bone that don't go all the way through (they're incomplete). Stress fractures are usually caused by use trauma.

INDEX

X

Z

INDEX